PROFITABLE SHEEP FARMING

M. McG. COOPER

C.B.E., B.Agr,Sc(NZ), B.Litt. (Oxon), Honorary D.Sc. (Massey),
F.R.S.E., F.R.A.S.E., Hon. Fellow, Wye College, University of London,
Formerly Dean of Agriculture now Emeritus Professor, University of
Newcastle upon Tyne

and

R. J. THOMAS

B.V.Sc., M.Sc (Bristol), PhD (Dunelm), M.R.C.V.S., Senior Lecturer in
Animal Health, Dean of Agriculture 1978–81, University of Newcastle
upon Tyne.

FARMING PRESS LIMITED
Wharfedale Road, Ipswich, Suffolk

First published 1965
Second edition 1971
Third (revised) edition 1975
Second impression 1976
Third impression 1977
Fourth (revised) edition 1979
Fifth edition 1982
Second impression 1983
Third impression (with amendments) 1986

COPYRIGHT © FARMING PRESS LTD 1982, 1986

Cooper, M McG.
   Profitable sheep farming. — 5th ed.
   1. Sheep — Great Britain
   I. Title   II. Thomas, R.J. (Robert John)
   636.3'00941   SF375.5.G7

ISBN 0 85236 117 3

Made and printed in Great Britain by
The Garden City Press Limited
Letchworth, Hertfordshire SG6 1JS

CONTENTS

PAGE

AUTHORS' PREFACE 11

**Production and Management
(M. McG. Cooper)**

1 INTERNATIONAL IMPORTANCE OF
SHEEP 13

*Differing production objectives—Sheep-meat production and con-
sumption—Wool*

2 SHEEP FARMING IN BRITAIN 24

*Historical aspects—Stratification of sheep production—Prospects
for expansion*

3 BREEDS AND CROSSES 39

*Classification—Terminal sire breeds—Longwool crossing
breeds—Ewe breeds—Cross-bred breeding ewes—Wool characteristics*

4 NUTRITION 54

*Body condition and its assessment—Energy and protein—Nutri-
tion of ewes before and after mating—Nutrition of ewes before and
after lambing—Feeding of lactating ewes—Nutrition of hill ewes*

5 FLOCK MANAGEMENT 69

*General considerations—Choice of breeding stock—Flock main-
tenance—Timing of lambing—Mating—Lambing—Systems of
grazing—Weaning*

6 INWINTERING OF EWES   88

Trials on University of Newcastle farms—Benefits of inwintering—Drawbacks and their alleviation—General management considerations

7 STORE LAMB FEEDING   99

Alternative of refrigerated storage—Short-keep stores—Long-keep stores—Indoor finishing of lambs—Returns from feeding stores

8 BREEDING, SELECTION AND RECORDING 110

Selection objectives—Strength of inheritance—Value of recording—New Zealand's 'Sheeplan'—Recording in Britain—Development of a breeding plan—Breeding plan for a ewe breed—Improvement of crossing breeds—Synthetic and imported breeds

9 FACTORS CONTROLLING PROFIT   127

General principles—Management priorities—Stocking intensity and nitrogen usage—Labour efficiency

Hygiene and Disease Control
(R. J. Thomas)

10 DISEASE CONTROL IN THE EWE   136

Non-infectious diseases: Pregnancy toxaemia—Lambing sickness—Hypomagnesaemia. Infectious diseases: Abortion—Border disease—Metritis—Mastitis

11 HEALTH AND DISEASE IN THE YOUNG 146
LAMB

Lamb survival—Coli bacillosis—Navel-ill—Swayback—Copper toxicity—White muscle disease—Coccidiosis—Clostridial diseases: Lamb dysentery—Entero-toxaemia—Braxy—Gas gangrene —Tetanus—Black disease—Control of clostridial infections

12 GENERAL FLOCK HEALTH PROBLEMS   157

Foot-rot—Orf—Pneumonia—Scrapie—Pining

13 PARASITIC DISEASES   167

External parasites: Mites—Lice and keds—Blowfly—Ticks and tick-borne diseases. Internal parasites: Tapeworms—Liver fluke—Roundworms

Appendices

I Sheep-meat in some EEC countries 185

II Body condition scoring 186

III Composition of main feeds 188

IV Specimen concentrate mixtures 189

INDEX 190

ILLUSTRATIONS AND TABLES

Figures PAGE
1 Lifetime performance of Border Leicester x 74
  Cheviot ewes mated as lambs or yearlings
2 Effect of age on size of lamb crop from Clun 76
  Forest ewes
3 Lactation curve of Border Leicester x Scottish 85
  Blackface ewes
4 Relationship of gross margins per hectare to fer- 132
  tiliser 'N' usage per grass hectare
5 Gross margins per ewe and per hectare in rela- 133
  tion to stocking rate
6 Typical roundworm life history 178
7 Nematodirus life history 179
8 Sources and levels of roundworm infection on 181
  pasture

Plates
1 Draft hill ewes brought on to easier land 31
2 Breeding ewes working for their living immedi- 56
  ately after weaning
3 Nucleus Clun Forest flock at Cockle Park 76
4 Outwintering can be expensive in terms of both 92
  labour and tractor use
5 Inwintering on straw–note the baffle wall and 97
  Swedish hay boxes
6 Site and method of vaccination 154
7 Sloping-sided foot-rot bath 160
8 Trimming feet in foot-rot control 161
9 Drenched lambs grazing clean pasture 182

Tables

1 Sheep production in selected countries, 1979   15
2 Sheep-meat as a proportion of total meat pro-   19
   duction in selected countries, 1979
3 Performance of mature ewes of terminal breeds   40
4 Performance of mature longwool crossing ewes   42
5 Performance of mature ewes   45
6 Performance of crossbred ewes in lowland flocks   49
7 Performance of commercial ewes in upland   51
   flocks
8 Lambing percentages of ewes in various body   58
   conditions at mating
9 Levels of feeding for twin-bearing ewes in late   63
   pregnancy
10 Lowland flock results by ewe type, 1980   72
11 Results from lamb-finishing enterprises, 1980–1   108
12 Physical and financial results per ewe for lowland   129
   and upland flocks, 1980
13 Percentage contributions to top third superiority   130
   in gross margins per hectare

AUTHORS' PREFACE

Revision of an existing publication is a much more difficult and possibly a less rewarding exercise than writing it in the first instance. That was our conclusion when we were faced with the task of updating a book that first saw the light of day seventeen years ago. Much has changed in the sheep industry since then, both economically and technically, and so we decided it was sensible to take the easier option of what is for all practical purposes a new version rather than attempting to make piecemeal alterations to an existing text. We have endeavoured, however, to preserve one feature of the original publication in so far as it was an attempt to stimulate creative thought in an industry that still retains a greater element of tradition than almost any other major branch of farming.

For many years during the post-war period of expansion and consolidation in British agriculture sheep production was not given a level of encouragement by the Government comparable with that enjoyed by other branches of production and for many farmers it was an act of faith for them to retain their sheep flocks. Their loyalty to sheep has been rewarded for the industry now has a buoyancy that sadly is lacking in many other farming activities. Numbers of breeding ewes have been increasing steadily in recent years to reach their highest level for at least half a century. There has been no suggestion of sheep meat quotas nor has there been any build up of surpluses. It is one of the few areas where production within the Community is well below the level of consumption. If France could be persuaded to accept lamb as a normal article of trade within the Community the scope for expansion would be enormous because of the comparative advantages Britain has over other members of the

Community so far as sheep production is concerned.

We have been greatly helped in this by the extremely valuable reports that are now being produced by the Sheep Improvement Services of the Meat and Livestock Commission that give the substance of hard facts rather than conjectures on which to base discussions of the industry's problems. The value of these reports in the preparation of this book is gratefully acknowledged.

We also wish to acknowledge the important contributions made by the two ladies who converted manuscript into typescript. Hilary Cooper had the more unenviable task, in as much as she had not only to handle the longer Production and Management section but also to contend with her husband's execrable longhand complicated by much crossing out and interspersion of second thoughts. Hers indeed was a labour of love. The Hygiene and Disease Control section was not without its problems but these were competently resolved by Mrs Irene Mills, and R.J.T. is particularly grateful to her.

We hope that this book will give its readers some measure of pleasure as well as a modicum of profit.

School of Agriculture   M. McG. COOPER
University of Newcastle upon Tyne   R. J. THOMAS

Chapter 1

INTERNATIONAL IMPORTANCE OF SHEEP

DIFFERING PRODUCTION OBJECTIVES

Sheep farming is possibly the most diverse branch of animal production. There are hundreds of races and sub-races of sheep, each with its particular environmental niche and economic function. At one end of the climatic scale there are the fat-tailed desert sheep which quite remarkably survive and thrive where other types of sheep would perish. At the other extreme there are mountain breeds like the Scottish Blackface that stand up to high precipitation, including heavy snowfalls, on acid hills with indifferent herbage and produce a reasonable lamb and a valuable fleece. In between there are the big mutton breeds like the Suffolk and Oxford, and diminutive breeds like the highly prolific Finnish Landrace and the hair-covered Barbados Blackbelly which is also prolific and takes the ram twice in the same year. There are specialist milk sheep like the East Friesland, specialist wool sheep like the Merino and triple-purpose sheep that produce milk, meat and wool. They can be owned by nomads or by peasants who never leave the locality where they were born. They can be kept in small groups where the owner knows each sheep by name or in huge flocks numbered in tens of thousands where sheep are mustered no more than three or four times in the year.

Numerically sheep in their various forms are second to cattle in order of importance among domesticated mammals. Obviously it is difficult to get an accurate account of numbers but in 1979 FAO estimated that there were 1,084 million sheep, 1,212 million cattle, 763 million pigs, 446 million goats, 131 million buffaloes and 62 million horses. There were twenty-seven countries that had estimated sheep populations in excess of 10 million and among these the leaders were USSR (143 m), Australia (134 m), China (95 m) and New Zealand (63 m). Because the alternative of pork is denied to Moslems, sheep have a special importance in several North African and Middle and Near East countries, in particular Turkey (44 m), Iran (34 m), Pakistan (24 m), Ethiopia (23 m), Afghanistan (23 m), Sudan (17 m) and Morocco (14 m). Mutton is also a favoured meat in India with a sheep population of 43 m. Not many years ago this was the approximate size of the United States' national flock but here numbers have declined steadily over the past half-century. The recorded total in 1979 was only 12·2 m as compared with 20·5 m at the beginning of the decade. Mutton is not a popular meat in the United States as New Zealand discovered in its endeavour to develop a market there for lightweight lamb.

In the hotter parts of the Middle East, and generally in tropical climates, hair as opposed to wool sheep predominate and this is understandable because a heavy fleece is a handicap in very high temperatures. On the other hand wool-producing sheep are favoured in temperate regions where wool not only has a protective function but also a considerable commercial value. The production of milk, which is mainly used for cheese, is also an important consideration in a number of countries. The relative importance of milk, meat and wool varies from country to country and this is illustrated by representative figures in Table 1.

New Zealand is in a class of its own in respect of the intensity of production of meat and wool but the production of milk is no more than a novelty undertaking of quite recent origin. Sheep farming is the country's major industry, and favoured dual-purpose breeds such as Romney, Corriedale and Perendale are characterised by heavy fleeces of uniform quality which are the result of many years of careful selection towards these ends. They are run in large flocks with a low labour input and they are almost exclusively grassland sheep. The best lowland pastures will support 15–20 breeding ewes per hectare but taking into account upland farms that also run the same breeds of sheep stocking intensity will average about

Table 1. Sheep production in selected countries, 1979

Country	Population million	Wool 000 tonne	Meat 000 tonne	Milk 000 tonne
Turkey	43·9	56·0	280	1,100
Iran	33·7	17·3	186	687
Spain	14·5	28·0	119	240
France	11·5	22·2	163	822
United Kingdom	30·0	48·7	227	—
Australia	134·4	705·8	491	—
New Zealand	62·9	320·6	505	—

nine ewes per hectare. Sheep are not folded as they are in many older countries and they are kept in fenced fields throughout the year. Despite an emphasis on lightweight lamb, exported to several markets, which results in an average carcass weight of only 15 kg, total output of sheep-meat is proportionately higher than that of any other country listed in Table 1 with the exception of France where the average slaughter weight is 21 kg and a proportion of total output of sheepmeat comes from store lambs imported mainly from Balkan countries and fattened in France.

Spain is in the same sort of latitude as New Zealand but its sheep industry is vastly different. In part this is due to differences in rainfall but there are other factors as well. Though flocks are sometimes large by European standards they are small as compared with those in the Antipodes and almost exclusively they are herded by day and are returned to folds at night. Seldom does one see sheep grazing unattended in fenced enclosures. Often flock owners have no land of their own, apart from the sites of their folds, but through long-standing arrangements with land-owners they are able to run their sheep on stubbles and similar sparse grazings, sometimes on otherwise waste land. Lambing often extends from autumn through to the spring and there is not the concentrated seasonal lambing characteristic of both Britain and New Zealand. Transhumance still survives but on a steadily decreasing scale, with herded sheep moving from dry lowlands where they winter to the comparative lushness of mountain pastures during the summer. Paradoxically Spain, which is the home of the Merino, the outstanding wool breed, pays relatively little attention to the fleece attributes of its 14 million sheep. Average fleece weight is less than half that in Australia or New Zealand. Output of sheep-meat, which is disproportionately low when compared with that of neighbouring France, reflects the Spanish taste for young fat-free lamb. Average slaughter weight, which is only 12 kg as compared with 20 kg in the United Kingdom, is influenced by the luxury demand for baby lamb, usually a by-product of milk production, which is often slaughtered at under six weeks of age.

Australia provides the best example of specialised wool production. With about one-eighth of the world's sheep she is responsible for more than a quarter of the output of greasy wool with the bulk of production coming from semi-arid land climatically not unlike much of southern Spain. Productivity of these dry-land pastures has been greatly enhanced by the introduction of subterranean clover which, ironically, originated in the Mediterranean region. The Merino is the favoured sheep because of the value of its wool and its adaptation to the prevailing environment, for it is essentially a dry-land sheep. It is prone to footrot under wet conditions and its fleece, unlike that of the Scottish Blackface or the Cheviot is not the sort that sheds rain. It is not happy with close confinement and the great open spaces of the Australian sheep stations are very much to its liking in that they suit the breed's instinct to travel in flocks. Also, it is an easy-care sheep and this is important where labour is at a premium as it is in Australia. In these respects the Australian Merino retains much of the characteristics of its Spanish antecedents, which were mainly transhumance sheep, but as a wool producer it is now a very different proposition thanks to generations of intense selection for higher fleece weights and uniform high-quality wool.

There is an irony in this achievement because there was once a widespread belief that the Merino would deteriorate if it was moved from its native environment. In spite of this belief Spanish authorities put an embargo on the export of live sheep to safeguard their valuable monopoly of producing the world's finest wool but this was broken in the eighteenth century by royal gifts to a number of crowned heads that included George III who established a flock at Kew. On the advice of Sir Joseph Banks, the famous biologist who was a botanist on Cook's 1769 voyage to the South Pacific, Merino rams were made available to English flock-owners with the intention of improving wool quality. However the Merino was not a success and in this sense it supported the Spanish view that it would not thrive away from its native land. It was a very different story when the Merino arrived in Australia about the turn of the century and today Australian Merinos not only produce fleeces that are up to three times the weight of clips from their Spanish counterparts but also the wool is generally of much higher quality.

The Australian preoccupation with wool is reflected in a comparatively low output of sheep-meat because total production is slightly less than that from neighbouring New Zealand where there is less than half the number of sheep. The Merino is not a good mutton breed in the sense that carcass conformation is poor and the flesh is a dark colour. Moreover it is run on land that is not suitable for fat sheep production while low fecundity and losses due to periodic droughts combine to limit the number of young sheep that can be slaughtered for meat. An appreciable proportion of Australian wool, particularly on poorer grazings, comes from wether sheep which do not suffer as ewes do from the physiological stresses of reproduction. At the end of their useful wool-producing lives the residual carcass value of the Merino wethers is practically limited to their contribution to pet foods or the manufacture of blood and bone manure. The same is true of old ewes. Australia does have a substantial fat-lamb industry but this is in the higher rainfall coastal belt and is based on mutton and dual-purpose breeds and their crosses.

In the semi-arid areas bordering the Mediterranean and in the Middle East the emphasis moves from wool to milk and meat production. Iran, for instance, with a quarter of the sheep population of Australia, produces only 17,300 tonne of raw wool which is less than one-fortieth of Australia's output. Proportionately meat production in Iran is about 50 per cent higher than it is in Australia and milk production, at 687,000 tonne, puts Iran in third place in the league of sheep milk producers.

SHEEP-MEAT PRODUCTION AND CONSUMPTION

The term lamb, when it is applied to sheep-meat, seems to have a rather flexible connotation, especially when it appears on menus, but more strictly it applies to animals that still have a full set of milk teeth. The description milk-fed or spring lamb should properly apply to younger animals that have been slaughtered directly off their mothers. Hogget mutton conventionally refers to animals approaching the two-tooth stage, i.e. where they have their first two permanent incisors which erupt at about twelve months. Carcasses from older animals come under the general category of mutton. Trade figures seldom distinguish the various kinds of sheep-meat and usually they all come under the one umbrella, 'mutton and lamb'.

In 1979 world production of mutton and lamb totalled 5·5 million tonne as compared with 45·4 m tonne for beef and veal, 52·8 m tonne for pigmeat and 28·0 m tonne for poultry meat. Taking into account all meat produced by domesticated animals sheep contributed only 4 per cent to a total output of 138 m tonne. Nevertheless there are several countries where sheep, for a variety of reasons ranging from religious considerations, dietary preferences and environmental conditions, are the principal meat animals. There are others where there is a limited demand for mutton and lamb and this is reflected in the production figures. Table 2 presents details of the balance of mutton and lamb in the totals of meat production in a number of representative countries.

The figures in Table 2 relate to output and not to consumption which is influenced by the levels of imports and exports. Of the listed countries Australia and New Zealand are both exporters of sheep-meat. The United Kingdom exports some lamb to the Continent but the amounts are small in relation to her imports which now mainly come from New Zealand. Of the other countries listed in Table 2 Iran is the only one with substantial imports of mutton and lamb.

New Zealand's export trade mainly consists of lightweight lamb in the 11–16 kg carcass range. For many years the favourite fat lamb sire has been the Southdown which not only produces small carcasses that are very acceptable in

Table 2. Sheep-meat as a proportion of total meat production in selected countries, 1979

Country	Total meat 000 tonne	Sheep-meat 000 tonne	Sheep-meat per cent
USA	25,908	133	0·5
Morocco	210	40	19
Afghanistan	217	99	46
Iran	628	186	30
Pakistan	680	125	18
Turkey	858	280	33
Germany	4,609	29	0·6
United Kingdom	2,928	227	8
France	5,281	163	3
Australia	2,975	491	16
New Zealand	1,075	505	47

Britain but has the added virtue that a high proportion of its lambs can be drafted fat from grass, either off their mothers or later following weaning. Lambs by large late-maturing rams generally require some supplementary feeding and this adds to costs of production. Traditionally New Zealand's main market has been the United Kingdom which, prior to entry into EEC, had a consumption of about 10 kg of mutton and lamb per head, which was substantially the highest level of any European country at that time. Both New Zealand and Australia, which was also a supplier of the United Kingdom market, have successfully diversified their outlets. Japan and the wealthy oil states of the Middle East, in particular Iran, have become important outlets for New Zealand and significant progress has been made in North American markets.

France, with an average of about 4 kg per head, which is less than half the level of intake in the United Kingdom, has the highest rate of sheep-meat consumption among the Continental members of EEC (see Appendix 1). The French have a liking for lamb and consumption could be appreciably higher if France was prepared, unreservedly, to accept imports from the United Kingdom where the prevailing farm-gate prices are substantially below those in France. Despite protracted negotiations France has resolutely opposed unrestricted entry of British lamb because it would spell ruin for her domestic producers who can survive only by means of prices which are prohibitively high by British standards.

Consumption of sheep-meat is at an appreciably lower level in Germany, Italy, Belgium and the Netherlands than it is in France. Beef and pork are preferred meats in Germany and lamb is only acceptable if it is tender and relatively free of fat. Italians have a liking for lamb and consumption there would be considerably greater if cost was not a limiting factor. As things stand it is a luxury meat and, rather as turkey used to have a role just as traditional Christmas fare in Britain, spring lamb has a special place in Italy at Easter. Unquestionably the United Kingdom, with its comparative advantages in sheep production, could develop a lucrative outlet in the Community by supplying a quality of carcass that has consumer appeal if it was backed by aggressive marketing.

WOOL

Spinning and weaving are practices of great antiquity, and wool has long been an important article in international trade. Much of the wealth of medieval England was based on wool and its export to the Continent. Many of the fine buildings in areas such as the Cotswolds which were formerly important for wool production are a heritage of this prosperity. The traditional seat of the Lord Chancellor when the House of Lords is in session is the Woolsack, and this again is a recognition of the economic importance of wool in the Middle Ages.

Much more recently wool has been important in the development of agriculture in the New World, particularly before the advent of rapid refrigerated transport. It has two important qualities in this connection. First, it has a relatively high value per unit of weight, especially in the case of fine wools; and secondly, it does not deteriorate in transit or storage, provided reasonable care is taken. In the first eighty years of the last century the only substantial export from Australia, apart from gold and other minerals, was Merino wool. Much of this was hauled by bullock teams from back-country stations to ports were it was then loaded on to sailing ships for the long haul around the Cape of Good Hope to Europe where it would usually be unloaded in good condition. Before the century was half-way through Australia had a deserved reputation for the quality of 'Botany Bay' wool. There were similar developments in other countries with natural grazing suitable for sheep where the economic prospects for other forms of land use were unfavourable, for instance on the South Africa veldt, the tussock grasslands of New Zealand, Patagonia, and the western rangelands of the United States. Merino sheep achieved a dominance on such land which went unchallenged until the advent of rapid ocean transport, combined with refrigeration, about 100 years ago, when there was a swing to dual-purpose sheep except where conditions did not permit more intensive farming. In New Zealand, for instance, the Merino is now confined to high-country holdings where other livestock would be at a severe disadvantage.

Though wool has been an important source of income in pastoral countries the lot of the sheep farmers has not always been a happy one because of price fluctuations. In one year prices could be reasonable and in the following year returns could be barely sufficient to cover the cost of shearing, packing and transport to the point of sale. The position was particularly grim during the early 1930s when strong wools sold for a few pence per pound and sometimes were virtually unsaleable while the position with fine wools was almost as bad. There was a feeling at that time, especially among sheep farmers, that demand would be so adversely affected by the growing popularity of man-made fibres that wool would never regain its former position. Fortunately for the industry, these fears have not been realised despite successes achieved by chemical engineers in the production of a wide variety of synthetic fibres. First the 1939–45 war stimulated demand because wool is the raw material of military uniforms, greatcoats and blankets. There was a second boost in 1950 during the Korean hostilities when wool prices temporarily reached an all-time high and, though they soon receded to a more normal level, prices have generally been reasonable since then.

In part this has been due to more sophisticated marketing

arrangements—sometimes with intervention buying—in the main exporting countries while the promotional activities of the producer-financed International Wool Secretariat have also helped to sustain demand. The Secretariat's Woolmark sign is now a well-recognised indicator of quality in the clothing trade. There have also been advances in the processing of wool; in particular the danger of woollen garments shrinking has now been greatly reduced. Above all, despite its best efforts, the synthetic fibre industry has only been partially successful in incorporating in its products some of the important natural attributes of wool, in particular resilience which not only contributes to shape retention in knitted and woven garments but also improves their insulation value because a springy yarn traps air between the fibres. Resilient wools with a high bulk per unit of weight factor, are especially important in the manufacture of warm, light-weight knitwear, blankets and rugs. Again, resilience is an important attribute of wools used in carpet manufacture to ensure that the pile does not show foot-marks. Another advantage of woollen garments over those made from synthetics is their capacity to absorb moisture and thereby minimise the discomfort arising from wearing wet clothing.

Another factor sustaining the demand for medium and fine wools has been their use with synthetics to produce fancy fabrics that have a special appeal for women. Fortunately for wool producers, women, unlike men, do not generally retain an affection for their old clothes and in an affluent society obsolescence comes very quickly in a woman's wardrobe.

There was a period about the middle fifites when it seemed, because of price advantages as well as their uniformity, that man-made fibres would take over from wool in the manufacture of all but the highest quality of carpets. This industry requires a strong wool with a high proportion of heavy fibres, such as that produced by the Scottish Blackface and similar upland breeds. Soon after the war there was a shortage of this type of wool because of the disturbed situation in North-East Asia which had been a main source of supply, and this deficiency was made good by a greater reliance on synthetic yarns. Synthetic carpets were extremely popular for a period because of their relative cheapness but it was soon realised that price is not the only consideration. The general experience has been that synthetic carpets quickly become shabby and in consequence they are not a satisfactory substitute for quality carpets made from wool. Fortunately for the farmers wool is now much more competitive. On the one hand the cost of crude oil, which is the principal raw material of the synthetic industry, has risen spectacularly and on the other there are now adequate supplies of suitable wool. In part this is due to the successful use of strong wool, such as that from the Romney, in blends that include traditional carpet wools. The supply of the latter wool has been enhanced by the development and growing popularity in New Zealand of a new breed, the Drysdale, that produces an ideal fleece for carpet manufacture.

A further attribute of wool is important in countries where the export of raw wool is a major source of foreign exchange. None of the main importing countries, apart from the United Kingdom, has a substantial sheep industry and so there is minimal pressure to give tariff protection to domestic producers. The EEC operates levies and quotas to limit and even bar importation of many products of temperate agriculture such as meat, dairy produce, grain and sugar but it maintains an open market for wool. It would not be in its interests or that of Japan, another important market, to do otherwise. In general it can be fairly safely concluded that the outlook for wool producers is a favourable one.

Chapter 2

SHEEP FARMING IN BRITAIN

HISTORICAL ASPECTS

Until the eighteenth century, when sheep achieved a new status in British agriculture, the principal emphasis had been on wool. There were no defined breeds as we know them today but local races whose characteristics were largely determined by the environments under which they had evolved. On heavy land, such as parts of Lincolnshire and the Midlands, the local sheep were large-framed slab-sided animals producing strong wool and lean stringy meat, while on the chalk and gravel soils of southern England they were smaller, fine-wooled sheep. Moorland sheep were small, rough-coated animals with many of the characteristics of modern hill breeds. Under the manorial system of agriculture the sheep were herded by day on common land and folded by night on the fallow shift of the three-course rotation that was standard on the arable land. Volunteer growth on these fallows provided keep but other considerations were security and the transfer of fertility from commons and waste land to the arable fields. It must have been a hard life for these medieval sheep because there were no supplementary crops to provide winter sustenance and cereals were too precious to be spared for animals. At least sheep, with their superior adaptation to difficult environments, fared rather better than cattle.

The lot of both sheep and cattle was greatly improved following the Restoration by the introduction of two new crops, turnips and clover, by Royalist landowners returning from exile on the Continent. It took some time for these crops to become popular because the restraints imposed by the customs and rules of open-field farming were impediments to

24

change but from the beginning of the eighteenth century there were important innovations and reforms. One was the Enclosure movement which created fields as we know them today that were under the control of the individual farmers who at last had some freedom of action. The second was the development of the Norfolk four-course rotation, popularised by the example and advocacy of an influential East Anglian land-owner, Lord Townshend, who earned through his efforts the sobriquet of 'Turnip' Townshend. In its most usual form the rotation was wheat—roots—barley—clover but there could be local variations with oats substituted for one of the cereals.

Grain yields were greatly increased by the rotation. The residual nitrogen from ploughing in the clover was utilised by the following wheat crop while the turnips, sown in drills, were a cleaning crop, both in respect of weeds and cereal diseases. More than this, the clover—usually sown with grass to be harvested as hay—and the turnips provided winter keep, particularly for yarded cattle, and the resulting farm-yard manure was returned to the arable land to enhance crop yields. The greatest contribution to the welfare and role of sheep in British farming came towards the end of the eighteenth century and it was made by another East Anglian landowner, Coke of Norfolk, later the Earl of Leicester, who folded sheep on a range of forage crops to build up fertility on the light and not very productive land on his estate. His so-called 'sheep shearings' at Holkham achieved international fame, so great was the success that he achieved, and as many as 600 people attended on these occasions. They witnessed how land should be farmed and how sheep could be managed with two sets of profits, one gained direct from the sheep and the other from the higher crop yields. His example was taken up by other landowners, many with home-farms where innovations were demonstrated to tenant farmers. The Holkham example had particular value for those farming the Downs and the Wolds and the light gravel soils of river valleys. Here folded sheep, as walking manure distributors, were the light-land substitute for the yarded cattle of stronger land and deservedly earned the reputation of having golden hooves.

There was another important contributor to this agricultural revolution which created the successful integration of livestock and crop production. He is Robert Bakewell (1725–95), a Leicestershire yeoman farmer who is unreservedly regarded as the father figure of livestock breeding in Britain. He was an exceptional man and he pioneered selection and mating procedures which anticipated later advances in the knowledge of heredity. The stage for his contribution had been set by enclosures and by the improvements in nutrition resulting from clover, sown pastures, roots and similar forage crops. Under the conditions imposed by common grazing a farmer had little scope for selecting sires and controlling mating. Moreover poor nutrition was a great leveller and it was not possible for farmers, limited to eye appraisal, to distinguish genetically superior stock. The only meaningful selection was that imposed by nature emphasising physical attributes which promoted survival rather than those favouring economically important traits such as growth rate, carcass conformation or fleece weight. With good nutrition it was at last possible for animals to realise more of their genetic potential and for breeders to distinguish the individuals best suited to their requirements.

Bakewell worked successfully with Shire horses, Longhorn cattle and Leicester sheep, but it was with sheep that he achieved his most meaningful results. In the space of about twenty years he transformed narrow, late-maturing sheep into very different animals with a spring of rib and a fattening propensity. A most important tool in this was progeny testing, and because numbers are important in animal improvement programmes (a topic which will be discussed in greater detail in a later chapter on breeding) he formed the Dishley Society, a group of farmers who co-operated with Bakewell in testing young sires. His sheep and the methods he adopted had their critics but there was no denying the impact that he had on the attitudes and practices of breeders not only in Britain but also abroad.

An important feature of the Dishley Leicesters was a uniformity which would have been in striking contrast to the prevailing heterogeneity of most of the contemporary races of sheep. It is not surprising, therefore, that Bakewell rams were widely used as improvers and it has been stated that most of the British Longwool breeds have Dishley blood in their ancestry. Certainly this is true of the Border Leicester and its offshoot the Hexham Leicester, and this is confirmed in letters written by Bakewell to George Calley, a Northumbrian farmer, which record movements of Dishley rams to the Borders and also much of Bakewell's philosophy.*

In the early part of the nineteenth century the emphasis in sheep breeding moved gradually from wool to meat as a consequence of a substantial increase in the human population combined with improved standards of living of the middle classes who were profiting from the Industrial Revolution and the expansion of international trade. The latter included increasing quantities of raw wool from the New World but as yet meat imports were excluded by time and distance.

As a consequence of these trends the initiative in sheep improvement moved from the Midlands to Sussex and several breeders, notably Ellman followed by Webb, developed a butcher's sheep from local stock once renowned for fine wool, which came to be known as the Southdown. Ellman was a butcher as well as a farmer and he recognised the needs of his trade which required a deep-fleshed, early-maturing animal with well-developed hind legs and a full loin. He succeeded admirably working with animals reared under favourable nutritional regimes to create the Southdown which soon earned the name of being an outstanding arable sheep. It was first used as a pure-bred—its cross-breeding role was to come later.

Just in the same way that the Dishley Leicester was used to improve other Longwool breeds the Southdown became important in the evolution and improvement of several other arable breeds that are now collectively described as Down sheep. Further to the west there was the Hampshire which later, with a second Southdown cross, created the Dorset Down. The Oxford, which is the biggest of the Down breeds, followed by the Suffolk and the Shropshire, all owe something to the Southdown, and now New Zealand has a breed

* See *Robert Bakewell*, H. C. Pawson (Crosby Lockwood, London, 1957).

called the South Suffolk which local breeders have created from a Southdown–Suffolk foundation.

These breeds held a firm place on light land during the period of high farming which lasted for about fifty years and terminated with the prolonged agricultural depression that started in the 1870s when grain from North America flooded into Europe. The period preceding this price collapse was in the nature of a high-water mark for sheep because not only were they profitable in their own right but, in the pattern of Coke's methods and advocacy, they were essential agents in profitable grain production, even on thin soils which were described as sheep and barley land. But first the bottom fell out of the grain market and then meat prices started to fail with the growth of imports of frozen lamb and beef from the Americas and Australasia. The consequence was that much light land, and also very heavy corn-land such as the lias and London clays, tumbled down to grass.

A new style of sheep farming gradually developed which was based on what came to be known as grass sheep to distinguish them from the arable breeds that had been favoured in combination with grain production. These included long-wooled lowland breeds such as the Romney, Devon Longwool, Leicester and Lincoln and also invaders into the Midland counties from the Welsh marches such as the Clun and the Kerry. In addition another class of ewe came increasingly into the picture as a fat-lamb mother, namely the first cross from Longwool rams, most commonly the Border Leicester, on ewes of upland and mountain breeds such as the North and South Country Cheviot, Scottish Blackface and Swaledale. The role of Down breeds altered in response to these developments. Generally they are not prolific breeds but they have a good carcass conformation combined with early maturity and their new function primarily became one of supplying rams to improve the carcass attributes of the terminal generation bred from prolific grassland ewes.

STRATIFICATION OF SHEEP PRODUCTION

The previously mentioned developments led to the organisation of Britain's sheep industry which essentially is based on a three-tier system that has evolved in response to natural conditions, economic pressures, the biological attributes of the different kinds of sheep, and advances in the basic technologies of sheep production, especially nutrition and disease control. A factor of cardinal importance in this context is that sheep are comparatively inefficient from both a biological and an economic viewpoint in the utilisation of the limited land resources of this country. Unlike New Zealand, despite the advantages of more prolific ewes and higher prices for fat lambs, it is not generally possible to devote good land to specialised sheep production because margins are usually insufficient to cover overheads and leave a satisfactory profit. Sheep are not in the same gross margin league as dairy cows and the same holds for cash crop production where land is suitable for this purpose. Sheep, in consequence, can only be supplementary to some other more profitable enterprise on lowland farms.

The situation is very different on hill farms, the first tier of the industry's stratification, where there is limited scope for much more than gradual improvements of grazing by such surface treatments as drainage, liming and phosphate application. Here, in the absence of differential subsidies or price support, hill breeds of sheep are the masters because of their adaptation to what is invariably a very rigorous environment. Here cattle, even hardy breeds like the Galloway or the Luing, are at a very considerable disadvantage as compared with the Swaledale, Scottish Blackface, Welsh Mountain and similar breeds. Among other attributes they are small and active and they fossick rather than graze as cattle do, utilising a wide range of herbage such as draw moss and heather when pasture is limited as it is for the greater part of the year on Britain's hills. Another physiological advantage comes from the appreciably shorter gestation and the shorter suckling period of sheep as compared with cattle. Another is the remarkable capacity that hill ewes have for producing viable lambs even though they have lost a high proportion of their body weight during gestation.

Despite these ecological advantages, cattle made the greater contribution to the substantial increase in productivity from hill and upland farms over the period from. 1950 to 1970 but this was due to political decisions originating with Lord Woolton's electoral promise at a time when meat rationing was still very strict, to increase the supply of 'good red meat'. One measure consisted of liberal headage subsidies for hill cows and their calves at weaning and there was also a succession of February price reviews that gave beef preferential treatment over lamb and mutton. At one stage direct subsidies, when compared on a livestock equivalence basis, showed a four-fold advantage for cattle and this coincided with a ten-year period when there were only derisory increases in the guaranteed prices for lamb. It did not create an encouraging climate for sheep production on either uplands or lowlands and it almost seemed that Government policy was designed to protect lamb imports in the interests of reciprocal trade.

Since accession to EEC the competitive position of sheep has improved enormously. Indeed a continuing worry is that there could be a substantial reduction of cattle in hill farming and this would be unfortunate not only because of the reduced output of store cattle but also because of the importance of cattle in the management of hill pastures where they act as mobile mowing machines to improve the balance of species in the interests of sheep. This function of cattle on hill sheep farms is recognised in many parts of the world and it was strongly advocated by those two notable leaders in the post-war resurgence of hill farming, Captain Bennett-Evans of Plynlimmon in Wales and Duncan Stewart on Ben Challum in Perthshire.

It was once usual for hill farmers to off-winter replacement ewe lambs on lowland farms but the number of farms prepared to take these sheep has declined and this has encouraged hill farmers to extend and improve their in-bye land. Some are in the fortunate position of having sufficient grazing and fodder crops to finish their wether lambs but more commonly these are sold as stores for fattening on lowland farms. This is one example of the valuable role of hill farming in the total fabric of British farming. Another is provided by the sale of cast-for-age hill ewes. Except when snowstorms have caused serious losses most hill farmers prefer to sell their ewes not later than the fresh full-mouth stage, that is just after they have reared their fourth crop of lambs. They still have a useful reproductive life ahead of them but with increasing age they become more vulnerable to the rigours of a life on the hill. Furthermore their value plummets once they have broken mouths so there are at least two good reasons for selling ewes before they are past their prime.

Occasionally hill farmers are in the fortunate position of having additional land where they can retain draft ewes until the end of their working days but more commonly they go to mainly stock-rearing farms rather than the more intensively farmed lowlands (Plate 1). These so-called marginal or upland farms constitute the second stage in the stratification of the industry. Draft ewes are usually mated to rams of Longwool breeds such as the Border Leicester with the resulting male lambs ending up as slaughter sheep that are either finished on their home farms or are sold as stores for finishing on lower ground. The ewe lambs are much too valuable as potential fat-lamb mothers to go for immediate slaughter and

PLATE 1
Draft ewes brought on to easier land for cross-bred lamb production.

Photo B. Tyrer

normally their working life is spent on lowland farms producing a terminal generation sired by rams of the specialised mutton breeds.

This is the final phase of stratification which in its entirety constitutes a very effective means of utilising diverse land resources. The true hills would have little more than an amenity value but for breeds like the Scottish Blackface that are adapted to difficult environments. Hill farmers must place an emphasis on a capacity to thrive and survive rather than on growth rate, prolificacy and carcass qualities; the use of Longwool rams not only improves the carcass attributes of the resulting half-bred lambs but, more importantly, because they are the principal product of the second stage of stratification, boosts fecundity in the females which also retain to an appreciable degree the highly developed mothering instincts of their hill ewe parentage. Moreover, lambs produced from these cross-bred ewes, even in a second generation away from the hill, are invariably very lively at birth and in this respect contrast with the sluggishness of newly born lambs of lowland breeds. This is an important factor in influencing survival during the inclement weather that so often characterises lambing. Finally, on the most expensive land, the terminal sire breeds make their special contribution of higher growth rates and improved conformation to maximise the value of the final lamb crop.

There is also a parallel system of two-tier stratification which features upland breeds like the Clun Forest, Radnor and North Country Cheviot that produce up to four crops of pure-bred lambs on marginal land and then move to lowland conditions to complete their working lifetime as fat-lamb mothers. This incidentally is the pattern of much of the sheep farming in New Zealand. Breeds like the Romney and Perendale are bred pure, usually for up to four crops on hill land where conditions are such that only a small proportion of lambs will be good enough for drafting fat off their mothers. They then move to lowland fat-lamb farms where the combination of better nutrition and earlier maturity derived from the use of Down rams, results in a higher proportion of milk-fat lambs.

The one serious drawback resulting from three-tier stratification is that it results in three sorts of slaughter lambs—those of the pure hill breeds, the first cross lambs of second-stage stratification and finally the double-cross lambs from the third stage. This variability in conformation is further compounded by the multiplicity of breeds and crosses. With more than forty pure breeds and many more two- and three-breed combinations and all the variations there are in husbandry and nutrition it is not surprising that there are frequent complaints of the lack of uniformity in home-produced lamb as compared with New Zealand imports. In the latter country there are fewer breeds, mostly with a similar conformation, producing export lambs. On top of this these lambs are seldom more than eight months old at slaughter and they are processed through large slaughterhouses where they are classified according to weight, conformation and finish and they are offered for sale according to well-established and meaningful grades. The bulk buyer for a chain of retail outlets can buy 10,000 New Zealand lamb carcasses of a given specification in the space of a few minutes on a telephone but it could take him a week to arrange the purchase of a similar quantity of home-grown lamb. It is little wonder that the meat trade, increasingly involved in centralised buying and chain distribution, is so often critical of the home product.

PROSPECTS FOR EXPANSION

Sheep are now relatively less important in British agriculture than they were fifty years ago. One can travel all day in East Anglia and see only an occasional flock of sheep and this in districts where sheep were once the lynch-pin in the maintenance of fertility in tillage farming. The industry suffered two serious body blows in the 1940s. The first was the large-scale ploughing out of permanent grass which was directed primarily to an increase in crops either for direct human consumption or for the support of milk production which, sensibly, had been given preferential war-time treatment as compared with other forms of animal production. The result was that sheep production during the War Committee years was largely concentrated on hill and marginal farms and this made the industry even more vulnerable to the second blow, the winter of 1947 which was the harshest for more than fifty years. It was estimated that at least 2 million sheep perished as a consequence of this prolonged period of blizzards and snow.

Throughout the fifites and sixties, as a result of a Government policy that has been mentioned briefly already, sheep farming became something of a Cinderella industry in the sense that a succession of annual price reviews offered little in the way of incentives to expand sheep production or to stimulate changes in technology comparable with those taking place in other branches of farming. Over the twenty-five years following the war average cereals yields rose by at least 50 per cent and the efficiency of food use in commercial egg production increased by at least 80 per cent. A totally new branch of poultry production in the shape of the broiler industry came into being to give poultry meat greater quantitative importance than lamb and mutton. The norm of two men on a 50 hectare dairy farm with forty milking cows averaging 3,500 litres moved up to eighty cows with an average of 5,000 litres. Meanwhile the sheep industry appeared to be moving backwards.

A report from the National Economic Development Committee in 1968 suggested that the only possibilities of expansion appeared to be on marginal land and it was not hopeful that the then current trend of lowland farmers going out of sheep would be reversed. This view was expressed long before there was any hint of vast surpluses of grain and milk products. It pointed out that it is comparatively easy for a lowland farmer with other land use opportunities to go out of sheep because most of the capital requirement is in the flock itself, an easily realisable investment, with very little in specialised buildings and equipment. In this sheep production differs from dairying where there is a considerable investment additional to land and livestock in specialised plant and buildings that have a limited redeployment value. It is not surprising that many farmers, even in such traditional sheep areas as the southern chalklands and the Scottish Borders, sold their breeding ewes and increased cereal acreages to adopt a simplified and, hopefully, a more profitable system of farming.

The flight from sheep was accelerated by the prevailing cult of measuring farming success in terms of gross margins without proper regard for other considerations such as safeguarding capital investments or maintaining the productive capacity of land. In the sixties only a well-above-average sheep enterprise boasted a gross margin of £20 per acre but under comparable circumstances cereals were at least double this figure. Many farmers on land formerly farmed on a basis of alternate husbandry had cause to regret selling livestock and redeploying their capital in wasting investments such as machinery in a system of continuous cereal growing. In many instances yields dropped alarmingly not only because of greater incidence of diseases and weeds but also because of a deterioration in soil structure which made cultivations more difficult and impaired crop growth.

An age-balanced herd or flock, particularly one that is maintained by home-bred replacements, is a hedge against inflation but no matter how well one maintains a tractor or a combine it eventually ends up as scrap and, with inflation, the cost of its replacement invariably seems to increase more rapidly than the level of the returns it generates. But a sheep flock, especially on arable land, cannot be justified purely on grounds of protecting capital investments or preserving soil structure. It must also provide a reasonable return on capital and other inputs. It is true that returns have improved substantially, relative to most other livestock enterprises, and in particular beef production, since entry into EEC and this is reflected by an appreciable increase in sheep numbers and a spectacular rise in the prices paid for breeding sheep. In 1979 there were 30 million sheep in the national flock which is 3 million more than the tally at the beginning of the decade. It cannot be said, however, that the greater buoyancy of sheep farming is attributable to any great increase in productivity or in its comparative efficiency.

One of the deepest thinkers among Britain's farmers, Oscar Colburn, who was responsible among other things for the development of Colbred sheep, makes the point that sheep production is still highly traditional in that it has failed to move with the times and there have been few innovations of the kind that has revolutionised other branches of farming.

His assessment is fair, for there have been no pioneers to blaze new trails as there have been in dairying, nor any meaningful changes in breeds such as those now in use in beef and milk production; and labour productivity has shown only marginal improvements over the past thirty years. Many still think that a 500-ewe flock is a full-time responsibility for a skilled shepherd but with wages at their present level he must either handle more sheep or do a lot of other jobs if he is to pull his weight in the economy of the farm.

There have been innovations that one felt had promise but their expected value has not been realised. The Colbred is one example for despite the success some farmers have had with Colbred crosses, particularly with the Clun, it has not made much ground. There have been several attempts to control the reproductive cycle and stimulate ovulation to increase the lamb crop and combine this manipulation of fecundity with artificial rearing but none of these ideas has yet proved to be commercially viable. Forward creep grazing appeared at first to be a promising means of intensifying fat lamb production along more conventional lines but now it has little more than academic interest. The biggest disappointment, however, is the industry's failure to realise the advantage to be gained from minimising the level of parasitic worm infection by a combination of field hygiene and strategic drenching that was clearly established over twenty years ago. The belief that a sheep's worst enemy is another sheep dies very hard and it is only comparatively recently that some vigorous extension work and convincing demonstrations by field advisory staff at the East of Scotland College of Agriculture have created an awareness of the importance of forward planning so that young lambs, which are highly vulnerable to worm infections, have the benefit of clean grazing. Their advice has started to take effect among sheep farmers in their region and this enlightenment is spreading over the Border into Northumberland. One can only hope that the new thinking will travel a little faster than Townshend's four-course rotation which is alleged to have spread at the rate of five miles a year from its base in Norfolk.

Factors operating against greater intensity in fat lamb production have included the standards set by fat stock graders and the high carcass weight before price penalties are imposed. It has been in farmers' short-term interests to produce over-fat carcasses which have been to the detriment of the demand for lamb.

Fortunately the absurd situation where farmers have been paid to produce something that consumers do not want has now come to an end with the development of more realistic grading standards. These have not yet had their full effect on practice for there are complaints that many farmers are continuing to market over-fat lambs because they are reluctant to depart from out-dated standards of what they consider to be a finished lamb. A full acceptance of the new grading objectives is important in several respects including safeguarding of the domestic market, and an enlargement of the Continental market which requires lean, light-weight carcasses. Another important consideration is that heavy, over-fat lambs are produced at the expense of carrying capacity and constitutes a waste of food that would be better used to produce additional lambs of the required quality.

We must think in terms of at least twelve ewes and twenty lambs to the hectare where sheep are the main grazing animal on good land if fat-lamb production is to be a competitive enterprise. The revised grading standards will help to make this goal feasible with the prolific breeding ewes that are freely available in this country. This intensity of ewe stocking does not compare with that attained on the better fat-lamb farms in New Zealand where a stocking rate of eighteen ewes per hectare is not uncommon but with the lower fecundity of New Zealand ewes only rarely will there be more than twenty lambs reared per hectare. Sired by Southdown rams, it is unlikely too that these lambs will average much more than 16 kg dressed weight to give an output of the order of 320 kg of lamb-meat per hectare. Twenty British lambs by a Suffolk ram out of Mule or Scottish Halfbred ewes, either sold off their mothers or within two months of weaning, should average at least 17–18 kg to give an output of about 350 kg of lamb-meat per hectare. Wool production will, however, have a very different complexion because of the much greater emphasis New Zealand farmers place on this product. Their ewes will clip about 5 kg of raw wool per head to give a total output of 90 kg per hectare as compared with a British figure of about 30 kg. But meat is a much more valuable product than wool under British conditions than it is in New Zealand and it is in British farming interests to put the emphasis on prolificacy and growth rather than on fleece weights.

Chapter 3

BREEDS AND CROSSES

CLASSIFICATION

The multiplicity of breeds and crosses in Britain has been getting even more excessive with recent introductions of Continental breeds and several attempts, some of them abortive, to develop new breeds. The likelihood, however, is that there will be some rationalisation with many local breeds retaining little more than a fancier's appeal, rather like Dexter cattle. Flock recording, one of the functions of the Sheep Improvement Services of the Meat and Livestock Commission, is at long last giving a basis for an objective assessment of breeds and crosses to replace what have been little more than generalised opinions, sometimes tinged with bias. The Commission distinguishes these three categories which are largely determined by their functions:

● Terminal Sire Breeds
● Longwool Crossing Breeds
● Ewe Breeds.

The last group is further subdivided into hill, upland and lowland breeds but these divisions are not as black and white in practice as they are on paper. For instance not all Suffolk-cross ewe lambs go for immediate slaughter because appreciable numbers are retained for breeding in an attempt to reduce flock replacement costs. Again the Clun, which is classified as an upland breed, is favoured by many lowland fat-lamb producers partly because they are able to breed their own flock replacements by mating about half the flock as pure-breds. Fortunately pure-bred Cluns generally have good carcasses though their growth rate will not compare with that of their Down-cross lambs.

TERMINAL SIRE BREEDS

These breeds are principally a source of tups for fat-lamb production because the emphasis in their selection has been placed on carcass qualities, especially rapid growth, early maturity and depth of fleshing, particularly in relation to the most valuable joints which are the leg and loin. They include the six Down breeds, the Ryeland which has some popularity abroad and three Continental breeds that have had some success, namely the Ile de France, the Texel and the Charollais. The first two after their earlier promise appear to have lost ground recently but the Charollais is increasing in popularity and understandably so because of its very well fleshed hindquarters. These are qualities that have a particular relevance in promoting lamb exports to the Continent.

MLC data are available for six breeds of this group and a summary of these is given in Table 3.

Table 3. Performance of mature ewes of terminal breeds

Breed	Mating weights kg	Lambs born per ewe	Birth weight* kg	Eight-week weight* kg
Dorset	74	1·41	4·75	18·6
Hampshire	72	1·41	4·6	20·1
Oxford	89	1·39	5·6	22·3
Southdown	55	1·49	3·65	16·0
Suffolk	83	1·71	5·15	21·6
Texel	79	1·76	5·0	22·0

* Unweighted average of twins and singles and males and females

The Oxford, now of minor importance, is the heaviest of the Down breeds and it is closely followed by the Suffolk, a more compact sheep which is the most popular breed in this group. The Southdown is much the lightest of the six breeds and the Dorset and the Hampshire occupy intermediate positions. The birth weight of lambs is closely related to the ewe size. The low weight of Southdown lambs suggests that rams of this breed could be useful mates for ewes bred to lamb at twelve months of age, in order to minimise the risks of distocia, but it must be remembered that Southdown-crosses are never more than lightweight lambs. The Dorset, which is second to the Suffolk in numerical importance, is probably a better proposition for such mating and with its somewhat neater head is possibly safer than the Hampshire. There is a good correlation between the eight-week weight of lambs and the milk yield of their dams. However differences between breeds in Table 3 must be treated with some reserve because these are accumulated flock records subject to differing seasonal and management effects. Comparatively small differences between breeds would only be meaningful if comparisons were made under the same environmental conditions. The slight superiority of the Texel over the other breeds in weight gain between birth and eight weeks could in this instance be due to better nutrition for it is still a novel and highly priced breed and on these grounds it is likely that it gets better than average treatment. The superior litter size of the Suffolk and Texel on the other hand probably has an inherent basis and the fact that there are now many Suffolk-cross ewes in fat-lamb flocks is further evidence that this superiority is inherent.

LONGWOOL CROSSING BREEDS

The principal function of breeds in this group is the supply of rams for mating to draft ewes from hill and upland breeds from the third category. Their main contributions are increased size, prolificacy and milk production in the resulting cross-bred ewes. The male cross-bred lambs go for slaughter, usually rather later in the season than Down-cross lambs, and often they are finished on fodder crops such as rape or swedes. There are six breeds in this group, namely the Border Leicester which is the most widely used; the increasingly popular Hexham or Bluefaced Leicester which is highly regarded in the north of England; the Oldenburg which is a newcomer originally from Germany; the Teeswater and

Wensleydale which are the sire breeds of the well-known Masham ewe, and lastly the Colbred which is now a breed of about twenty-five years' standing. The four breeds used in the Colbred's foundation are the Border Leicester, the Clun, Dorset Horn and an imported milch breed, the East Friesland. It is a tailor-made breed specifically designed to provide sires for cross-breeding and in this respect it differs from the other five which were initially important in their own right as pure-bred sheep but which subsequently achieved their cross-breeding role.

The Bluefaced Leicester, which can by no means be described as a handsome sheep, though it is undoubtedly a very useful one, is an off-shoot of the Border Leicester and was once a poor man's sheep. The modern Border Leicester is a white-faced sheep but earlier in its history there was a proportion that had face pigmentation that was probably inherited from Bakewell's Dishley sheep. The reduced sale value of these pigmented sheep appealed to the small-scale farmers of Hexhamshire who could not afford the higher prices realised by white-faced sheep.

The Bluefaced Leicester has the reputation of being Britain's most prolific breed and this is supported by the figures in Table 4. The lower birth weight of its lambs as compared with Border Leicester lambs (5·1 kg as against 5·8 kg) is explained by the higher proportion of multiple

Table 4. Performance of mature longwool crossing ewes

Breed	Body weight at mating kg	Lambs born per ewe	Birth weight* kg	Eight-week weight kg
Bluefaced Leicester	86	2·24	5·1	22·7
Border Leicester	83	1·86	,5·8	22·8
British Oldenburg	70	1·70	5·1	20·8
Teeswater	86	2·15	5·9	21·4
Wensleydale	103	1·79	6·2	23·75

* Unweighted average

births. Its eight-week lamb weights are practically identical with those of the Border Leicester despite this initial disadvantage and the greater competition for mothers' milk arising from the higher proportion of twins (1·99 living lambs recorded for the Bluefaced Leicester compared with 1·63 lambs for the Border Leicester).

One cannot say that this apparent superiority is due to blue pigmentation or preferential treatment. If anything, because it is a fashionable breed, the likelihood is that the Border Leicester has at least as good an environment as its counterpart. Possibly because the Bluefaced Leicester was originally a small farmer's sheep greater emphasis in selection has been placed on reproductive efficiency to give present-day sheep an advantage in this respect. Certainly it has a very high reputation in the North of England as a mate for draft Blackface and Swaledale ewes and the resulting sheep known as Mules have a well-deserved reputation as fat-lamb mothers. Within the past five years there has been a mushroom growth of so-called Welsh Mules which are the product of Bluefaced Leicester rams on draft Welsh ewes. Limited performance data now available for this cross suggest that it is at least as good and possibly a better mother than the Welsh Halfbred sired by the Border Leicester which now has a well-established place in Wales as it has had for many years in Scotland. The Scottish Halfbred which is the product of the Border Leicester on either the North Country or the South Country Cheviot has long been the most highly-regarded of all cross-bred ewes if price can be taken as the criterion of merit. The progeny of the Border Leicester on the Scottish Blackface is known as the Greyface. It differs only slightly from the Mule in appearance, the main feature being the darker face colouring of the latter cross which earns a premium in the estimation of most north of England sheep farmers. Either cross constitutes a fat-lamb mother of very high merit as later figures will show.

Further south the Masham, which is usually sired by the Teeswater and occasionally by the Wensleydale, is popular with Yorkshire farmers. Especially if it has a Wensleydale sire the Masham is a taller sheep than the Mule which is to be expected because the Wensleydale is the biggest of the British sheep breeds. Neither it nor the Teeswater is numerically important and certainly they are not in the same league as the two Leicesters. The Oldenburg was imported largely because it had a close resemblance to the Romney but with one important difference, namely it is more fecund. The Romney has many virtues but prolificacy is not one of them. The Oldenburg has mainly been used on the Romney to produce a halfbred that has had only a limited farmer appeal.

EWE BREEDS

Table 5 gives performance data for the three categories of ewe breeds. The list is by no means complete. Those omitted, because there are no available data, include the Lonk, Herdwick, Rough Fell, Dartmoor, Beulah, Hardy Specklefaced and the South Country Cheviot. It is important to remember that data for the three hill breeds derive largely from tup-breeding farms where particular care is taken in flock management. They cannot therefore be regarded as being generally representative because on most commercial hill farms with less intensive shepherding it would not be possible to record lamb weights. This reservation will apply also, but in a lesser degree, to the other two breed categories, but the figures do show a breed's capability under good conditions.

The Scottish Blackface is the heaviest of the three hill breeds and the Welsh is the lightest. There are types within breeds which show distinct differences, for instance the Lanarkshire Blackface is considerably bigger than the Galloway type which is better suited to harder country, while there are also divisions in the Welsh Mountain breed. Under commercial conditions in the north of England there are many hill flocks with mixed Blackface and Swaledale breeding. It is generally considered that the latter is better adapted to hard conditions than the Blackface which has the larger and more valuable lambs and the heavier fleece. In order to get the best compromise it is not uncommon for farmers to use tups of one breed for several generations and then follow with rams of the other breed. There is some experimental evidence to support this practice. A study undertaken by the Animal Breeding

Table 5. Performance of mature ewes

Breed	Mating weight kg	Total lambs per ewe	Birth weight* kg	Eight-week weight* kg
Hill				
Scottish				
·Blackface	54	1·46	3·8	17·3
Swaledale	48	1·49	3·4	18·2
Welsh				
Mountain	35	1·18	2·9	13·5
Upland				
Clun				
Forest	60	1·72	4·3	18·1
Derbyshire				
Gritstone	63	1·45	4·9	18·6
Devon				
Closewool	69	1·45	3·9	19·8
Kerry				
Hill	61	1·60	4·9	18·3
North				
Cheviot	73	1·81	5·4	22·1
Radnor	51	1·44	3·9	17·1
Whitefaced				
Dartmoor	54	1·42	n.a.	17·8
Lowland				
Devon				
Longwool	89	1·54	n.a.	21·2
Dorset				
Horn	72	1·57	4·3	19·3
English				
Leicester	94	1·51	5·8	19·6
Jacob	53	1·94	4·0	14·5
Romney				
Marsh	71	1·38	4·7	19·8
Lincoln	91	1·48	4·4	21·5
Llanwenog	50	1·80	4·6	16·3
Lleyn	53	1·96	4·0	16·8
Poll				
Dorset	74	1·67	4·3	18·9
Wiltshire				
Horn	67	1·71	4·4	21·7

* Unweighted average   n.a. Not available

Research Organisation has indicated a 10 per cent superiority in reproductive efficiency for Blackface x Swaledale ewes over the average of the two parent breeds of the cross. This suggests that hill farmers can expect a useful response, attributable to hybrid vigour, when they make this outcross. A characteristic of both Swaledale and Welsh sheep is the relatively rapid growth that their lambs make while suckling. This is confirmed by the figures in Table 5. Though Welsh lambs average only 13·5 kg at eight weeks, by far the lowest figure recorded for any breed, nevertheless this is 39 per cent of the weight of their dams at mating. The Swaledale with a figure of 38 per cent is equally impressive because here the proportion of twins is greater. The Blackface figure on the other hand is only 32 per cent but this is also high compared with lowland breeds. A good flow of milk from the dam and a relatively rapid growth in the lamb are important survival traits in difficult environments. This characteristic is not strongly expressed in lowland sheep, even those with a litter size comparable to that of the Swaledale. For instance Lincoln lambs at eight weeks achieve only 24 per cent of their dam's weight at mating and this is also the Devon Longwool figure. The Romney Marsh is slightly higher at 28 per cent.

The North Country Cheviot compares favourably with other members of its group with an average litter size of 1·81 which is only approached by the Clun Forest with 1·72. The eight-week lamb weight of 22·1 kg, despite a high rate of twinning, is the highest recorded for any breed, regardless of category and it is just over 30 per cent of the ewe's average weight at mating. On this evidence of mothering ability it may seem surprising that North Country Cheviots mated to Down breeds are not as popular as Clun ewes, which are widely distributed south of Yorkshire, for intensive fat-lamb production. There is an explanation, namely that the logical mate for draft Cheviots is not a Suffolk but a Border Leicester to produce the North Country Halfbred which is the Rolls-Royce among cross-breds.

None of the lowland breeds apart from the Dorset Horn, the Polled Dorset and the Jacob has much more than a local appeal. The Romney still has some support north of the Thames but its most important base is in the south-east corner of England where it originated. It is a very different sheep in both body conformation and fleece characteristics from the New Zealand Romney which is a blockier sheep with much more uniform and heavier fleece. The export of Romney rams to New Zealand or the Argentine, another country that has favoured this breed, is now very much a thing of the past. The same is true for the Lincoln and the English Leicester. This last breed was once important on wold farms in Yorkshire but its position has been eroded by the Masham and other more prolific cross-breds. A few years ago the Jacob, which is claimed to be the oldest breed in Britain, was a comparative novelty but this is no longer the case, for now it is a rather picturesque and fairly common hobby sheep favoured by people whose living comes from some other source than these sheep. It is claimed that the Jacob's meat has a gourmet's appeal but it is unlikely that this testimony to its excellence or its remarkably high prolificacy will ever make the breed truly commercial. It is rather surprising that the Lleyn and the Llanwenog, which are among the most prolific of Britain's lowland breeds, have not made some impact outside of west Wales where they are held in high regard. If the Lleyn had originated in East Wales, as the Clun did the chances are that it would have exceeded this breed in its popularity in England.

The Dorset Horn and its derivative the Polled Dorset have a unique role in that they are the only British breeds that regularly produce out-of-season lambs. As a general rule, the further a breed's place of origin is from the Equator the more restricted is its breeding season. This is a logical outcome of pressures from natural selection operating where there is a marked seasonality of food supply. In cold temperate climates lambs that are dropped in the spring have a greater chance of survival because their vulnerable early stage coincides with a period of adequate food supplies. Scottish Blackface ewes, evolved under these conditions, will seldom take the ram in August and September but the Suffolk, evolved in much more favourable environments, comes in season very rapidly at this time. The Merino originating in a Mediterranean climate characterised by autumn growth and very dry summers, regularly conceives in the spring and drops its lambs in the back-end of the year when conditions are relatively favourable. Nearer the Equator, breeds like the Egyptian Ossimi and the Persian Blackhead take the ram throughout the year and quite often while they are suckling, a trait that the two Dorsets do not regularly manifest. There is a tailor-made breed in South Africa called the Dorper which is an amalgam of the Dorset Horn with its extended breeding season and reasonably good carcass attributes and the Persian Blackhead's capacity to conceive while suckling with the aim of producing two crops of lambs in a year. The best one can hope from the Dorset Horn in Britain are three reasonably good crops of lambs in two years but attempts to achieve this can lead to management complications. Possibly the best, and certainly the simplest, arrangement is a concentrated effort to cater for the luxury Christmas and Easter trade in young milk-fed lamb by always lambing in the back-end of the year. In Australia the two Dorsets are widely used as fat lamb sires in matings with other breeds but this is not their usual function in Britain.

There is one breed not listed in Table 5 that deserves some mention not because it has a commercial future as a pure breed but because of its contribution in the evolution of new breeds and crosses. It is the highly prolific Finnish Landrace which as a mature sheep will regularly produce litters of three, four and even five lambs which, despite their low birth weights, have an amazing will to live. The breed is characteristically very small and it has a poor body conformation and little merit as a wool producer. Its value rests in its prolificacy and in another feature, its extreme docility. Even the first cross with a lively breed like the Clun, that is constantly on the move, is a very relaxed sheep and this could be important with intensive systems of production where sheep are closely confined. The breed was used to establish the Cadzow Improver, now defunct, which was a first cross of the Finn with the Dorset Horn. The intention was to produce an alternative to the Border Leicester that would sire prolific cross-bred ewes with an extended breeding season. This concept has not been abandoned entirely because the Animal Breeding Research Organisation has produced a sire-line containing Finnish blood which is currently under comparative farm tests.

CROSS-BRED BREEDING EWES

The Sheep Improvement Services of MLC have also collected performance data for the principal classes of cross-bred ewe under both upland and lowland conditions. Again, as in the case of pure-breds, there is the reservation that these do not have the validity of performances recorded under standard conditions but they are still valuable because they cover six years of collection over a wide spectrum of management so that any major differences are likely to have a practical significance. Table 6 is a summary of the performance of cross-bred ewes in lowland flocks and Table 7 covers corresponding records for commercial upland flocks. It also includes records for some pure-bred ewes under upland conditions.

There does not appear to be any meaningful difference

Table 6. Performance of cross-bred ewes in lowland flocks

			Per 100 ewes put to ram			
Breeding	Mature weight kg	Litter size	Ewes lambing	Total lambs born	Lambs reared	Lambs lost
Masham	71	1·79	92·8	174	151	23
Greyface	70	1·79	93·5	175	150	25
Mule	73	1·78	93·2	177	151	26
Welsh H/B	58	1·56	93·6	154	136	18
Scottish H/B	77	1·76	92·6	175	149	26
Romney H/B	76	1·38	93·1	137	119	18
Suffolk x Clun	72	1·52	93·0	148	128	20
Suffolk x Romney H/B	80	1·48	92·3	145	121	24
Suffolk x Mule	77	1·79	92·8	178	146	32
Suffolk x Scottish H/B	80	1·70	91·3	166	137	29
Suffolk x Welsh H/B	69	1·55	92·8	153	132	21

Note: Litter size refers to live lambs born per ewe

between the various crosses in respect of proportion of ewes lambing except that the four triple crosses with a Suffolk top cross are slightly inferior to their respective double crosses. However, this has not adversely affected their overall lambing performance which for all practical purposes is as good as that of the two-way crosses. A disturbing feature is that approximately 7 per cent of all ewes have failed to produce lambs either because of death during pregnancy, failure to breed or abortion. More alarming is the high mortality rate of lambs from birth to rearing. This is probably a better than average picture because the likelihood is that recorded flocks have higher than average management. Losses, both at birth and subsequently, are lower for the Welsh and Romney Halfbreds and the Suffolk x Cluns than they are for the more prolific crosses but this is insufficient to compensate for their smaller drop of lambs. However, for anybody subscribing to the New Zealand philosophy of minimal care sheep to reduce shepherding costs these less fecund crosses may have some appeal but the opposite viewpoint would seem to have greater validity, namely that the higher value of the lamb crop in Britain justifies the use of highly fecund ewes combined with high management standards, particularly at lambing. Twenty extra lambs reared per 100 ewes mated on a lowland farm creates a lot of additional income at very little additional cost.

There are of course factors other than prolificacy, mothering performance, and ease of management that determine the value of a cross-bred breeding ewe. A choice between Mule and Scottish Halfbred ewes can be influenced by the quality of a farmer's pastures and his lamb disposal policy. There is no question that Down-cross lambs out of Scottish Halfbred ewes are more attractive looking than corresponding lambs out of Mule ewes when they are sold on the hoof but when they are on the hook there is very little difference. Generally it is accepted that Mule and Greyface ewes are more easily wintered and are, to use an expressive Spanish idiom 'more rustic' than the Halfbred. It is noteworthy in this connection that Scottish Halfbreds do not feature in Table 7 which records performance in upland flocks that have to endure a less kindly environment than that of the lowland flocks.

The two most prolific cross-breds in recorded upland flocks are the Greyface and the Suffolk x Mule but a lot of the advantage arising from larger litter sizes is lost through higher mortalities between birth and weaning. This is a normal expectation but nevertheless it underlines the importance of good nutrition for ewes before and immediately after lambing and the provision of shelter for new-born lambs in the management of prolific flocks. Apart from this point it does not appear that any of the listed sheep has a superior adaptation to upland conditions. For instance the proportion of ewes lambing is virtually the same for each breed and cross- and peri-natal lamb mortalities differ only very slightly.

Table 7. Performance of commercial ewes in upland flocks

			Per 100 ewes put to rams			
Breed	Weight kg	Litter size	Ewes lambing	Total lambs born	Lambs reared	Lambs lost
Welsh H/B	56	1·54	92·9	152	133	19
Greyface	68	1·71	92·9	171	146	25
Suffolk x Mule	75	1·68	92·2	166	138	28
Suffolk x Clun	74	1·51	92·0	148	127	21
Suffolk x Blackface	68	1·44	92·6	141	123	18
Suffolk x Welsh H/B	66	1·55	92·4	151	129	22
Clun*	61	1·52	92·9	150	131	19
Beulah*	52	1·38	93·2	137	119	18
Scottish Blackface*	52	1·40	92·9	139	120	19
Hardy Specklefaced*	50	1·36	93·2	135	117	18
Swaledale*	50	1·41	92·6	139	121	18
Welsh Mountain*	38	1·23	93·2	121	106	15

Note: Litter size refers to live lambs per ewe
* Purebred flocks

WOOL CHARACTERISTICS

Despite efforts by the Wool Board to persuade sheep farmers to pay more attention to the production and handling of wool it is, in contrast to Antipodean standards, a neglected commodity in Britain. This is understandable when one examines the contribution that wool makes to total flock income. Over the four years 1975–8 the average return per ewe from the sale of lambs from lowland flocks costed by MLC was £28·37 as compared with £2·56 for wool which is a factor of 11 to 1. The ratio is even less favourable for wool in upland flocks where, over the same period, returns averaged £24·63 for lamb sales and the sale of wool averaged only £1·82.

One cannot expect British breeders to place much emphasis on wool under these circumstances. Inevitably if selection pressure is directed towards heavier fleece weights in a particular breed there will be some deterioration in other traits. This was New Zealand's experience with Romney sheep as a consequence of breeders' efforts to produce heavier clips. The muffle-faced sheep they bred, with well-woolled points and bellies, were such indifferent mothers that there had to be a switch back to open-faced sheep more like the English Romney. Though it does not make sense for British flockmasters to sacrifice reproductive efficiency in any quest for heavier fleeces this does not mean they should not aim for greater uniformity and also ensure that wool is offered for sale in the best possible condition.

The most striking feature of British wools is their extreme diversity. At one end of the scale there are the hill breeds with hairy fleeces averaging about 2 kg per head. In the case of the Welsh Mountain the typical fleece of a breeding ewe is no more than a kilogram. At the other extreme there are breeds like the Lincoln and the Wensleydale where well-managed flocks will average about 6 kg of wool which is virtually free of the kemps and hairy fibres that characterise the fleeces of many hill and upland breeds. There are also big differences in staple length, in fibre diameter, crimp, lustre and spinning qualities.

Southdowns have the finest wools but in common with that of the other Down breeds it has a short staple. At the other end of the scale the Leicesters, the Lincoln, Teeswater and Wensleydale have coarse, long-stapled wools. Breeds like the Clun and Cheviot are intermediate both in fineness and in staple length. There are also very big differences in handle. Down wools, despite the fineness of their fibres, tend to have a harsh handle and in the wool trade are considered to be more suitable for hosiery than for fine fabrics. Cheviot wool has a smoother handle and another important trait, resilience, which is of importance in the manufacture of lightweight garments, especially knitwear. Romneys produce a soft-handling medium-lustre wool and ideally these fleeces should be free of hairy or medullated fibres, but in practice a high proportion of English Romneys, as well as other Longwool breeds, have hairy britches which are readily recognised by their rather chalky appearance as compared with the pure wool of the shoulder region. Though hairiness is a fault in the majority of the longwool breeds it is a valuable trait in carpet wools. It is one of the most interesting features of the wool trade that every kind seems to have a special value for some particular purpose, be it the manufacture of felts, carpets, moquettes, blankets, tweeds, knitwear or worsteds. The wool trade, however, likes to have fairly uniform wool throughout the fleece rather than the variability characteristic of so much fleece wool. The most that the Wool Board can hope for in Britain is greater attention to uniformity, a lower incidence of 'tender' wools as a consequence of nutritional stress, and a reduction in faults due to thoughtless management practices such as bloom dipping, use of paints instead of acceptable marking materials, and the careless packing of fleeces. Any appreciable increase in wool production will only come from an expanded sheep population.

Chapter 4

NUTRITION

There was a saying that half the breeding goes down the throat—in other words that genetic expression is largely determined by nutrition. This is exemplified by stock in pedigree herds and flocks where returns come mainly from the sale of breeding stock but there is a world of difference between the nutrition of these animals and that of the ordinary run of stock on which commercial farmers depend for a living. Here it is a matter of establishing levels of nutrition that give an optimal economic response, for the law of diminishing returns applies just as much in livestock feeding as it does in fertiliser practice.

But the determination of economic feeding levels is not a simple exercise and this is especially so with sheep. First there is the range of conditions under which sheep are kept, the varying capabilities of the different breeds and crosses, and differing requirements at various stages of growth. Then there are the necessary adjustments to meet metabolic needs at different stages of the reproductive cycle. The task is further complicated through sheep being primarily grazing rather than trough-fed animals while herbage, their principal source of nutriment, varies both qualitatively and quantitatively over the year. On top of this there is no direct method for establishing nutrient intakes under farm conditions. There is a further complication, common to all grazing animals, and that is the risk of parasitic worm infections. This is of particular concern with young animals that have not developed any measure of age immunity. It is true that we now have a range of effective drugs for the control of specific parasites, while a better understanding of their biology enables farmers to take the necessary management precautions to minimise the

54

challenge of infections. Nevertheless control measures introduce another complication in planning the best use of grazing.

Fortunately trace element deficiencies, another cause of ill-thrift, particularly in ruminants, is no longer the problem it was before advances in micro-analysis established that certain elements, present in only minute amounts in feedingstuffs, are essential to normal health. Cobalt, selenium and copper are three essential trace elements, and areas in Britain where they are deficient are now well recognised so that farmers are able to take precautionary measures. Some minor elements can also be present in herbage in excessive amounts to the point that they constitute a health hazard. Molybdenum in 'teart' pastures of South-west England is one example and its effect is to induce copper deficiency which causes 'swayback' in lambs. But copper in excess is also dangerous. Even with quite small quantities in the diet it will gradually accumulate in the liver until fatal levels are reached. Sheep are at risk grazing in orchards where copper sulphate sprays have been used, and so are sheep on pastures dressed with pig slurry where a copper supplement has been included in the diet. Copper should never be included in mineral supplements because of this danger.

A further complication is provided by the so-called metabolism diseases hypocalcaemia and hypomagnesaemia, more commonly known, respectively, as milk fever and grass staggers. Hypomagnesaemia fortunately is not a widespread problem in sheep, but it sometimes occurs in old ewes with twins on herbage deficient in magnesium. It seems that increasing age impairs an animal's ability to mobilise its reserves of magnesium to maintain the necessary titre of this mineral in the blood. A luxury uptake of potassium at the expense of magnesium by quick-growing pasture in the early spring is also associated with low levels of blood magnesium. It is wise to delay potash dressings until the summer when clover, which is a rich source of dietary magnesium, is making a significant contribution. Cereal grazing in the early spring, particularly when the crop is starting to make active growth, has been associated with deaths attributed to hypomagnesaemia. It is advisable when ewes with lambs at foot graze proud wheat to have a grass run-off and to provide a concentrate fortified with calcined magnesite.

BODY CONDITION AND ITS ASSESSMENT

The condition of grazing livestock can deteriorate very quickly and it is one of the skills of a competent stockman that he is able to read the signs at an early stage in order to take the necessary corrective steps before there are serious set-backs. If sheep are on their feet for most of the day restlessly looking for food, instead of steadily grazing to repletion and then lying down to ruminate, one can be very certain that they are underfed. This should never happen with lambs which should be allowed to grow without any serious checks, while there are only two periods of the year where breeding ewes can reasonably be expected to work for their living (Plate 2) without serious detriment to their total performance. One is immediately after weaning and the other is the middle third of

PLATE 2
Between weaning and flushing, breeding ewes can be made to work for their living.

Photo University of Newcastle

pregnancy but even here it is wise to give preferential treatment to two-tooth or older ewes in low condition. Invariably in the absence of any disease or dentition factor it is the ewes which have done an honest job by rearing two good lambs that are likely to be in low condition and therefore deserving of special attention.

It is relatively easy to make a visual assessment of the body condition of cattle but the task with sheep is more difficult because of the masking effects of the fleece, except at the extremes of emaciation and gross fatness. Consequently handling is necessary if one is to get some more satisfactory estimate of condition be it poor, store, forward store or fat. But these are not precise terms and their value can differ with farming conditions; for instance a hill shepherd might describe as fat lambs what a lowland shepherd would regard as being no more than forward stores. In order to give greater precision to condition rating Australian research workers have devised a numerical system of scoring with a scale ranging from 0 to 5 that is described in Appendix II. A zero rating refers to extremely emaciated sheep that are little more than skin and bone and are near to the point of death while sheep in Grade 5 have lumps of fat over the tail-head and it is impossible by handling to detect the vertebral processes in the loin region. These are extremes and in practice a sheep farmer will be more concerned with sheep in the 2–4 condition range, particularly with breeding ewes to ensure that they are sufficiently fit to do the job expected of them while making the best use of food.

Condition rating, it is stressed, is not a scientist's gimmick but a valuable tool for improving economic performance. A sheep-farmer has to steer a safe course between the Scylla of under-feeding and the Charybdis of wasteful over-feeding, and the periodic sorting of ewes on a basis of their condition with reference to their immediate physiological needs is likely to become a standard management procedure.

There can be no doubt from the evidence in Table 8 that body condition has a profound effect on the drop of lambs even to the point that the size of the lamb crop could be an embarrassment under hill conditions where there may be insufficient good grazing to support a high proportion of ewes

Table 8. Lambing percentages of ewes in various body conditions at mating*

Type of ewe	Body condition score at mating						
	1	1½	2	2½	3	3½	4
Hill ewes							
Scottish Blackface	—	79	—	—	162	—	—
Hill Gritstone	—	—	75	103	119	109	—
Welsh Mountain	60	65	105	116	123	—	—
Swaledale	—	78	133	140	156	—	—
Lowland							
Gritstone (lowland)	—	—	—	132	154	173	—
Masham	—	—	—	167	181	215	—
Mule	—	—	149	166	178	194	192
Greyface	—	—	147	163	176	189	184
Welsh Halfbred	—	126	139	150	164	172	—
Scottish Halfbred	—	—	148	170	183	217	202

* Reprinted from *Feeding the Ewe*, published by Meat and Livestock Commission

with twins. There is some suggestion, too, that there can be a diminution of the lamb crop if ewes are mated in a very high condition and this is not surprising because such ewes are somewhat more vulnerable in pregnancy than their leaner mates if there is any severe restriction of food intake. Moreover over-fat ewes at any stage of the year constitute a waste of food.

It is not enough to use body score as a guide for preparing ewes for mating. It must be used subsequently to sort ewes for differential feeding as well as to provide a check on the general welfare of the flock. Merely weighing sheep is not enough because ewes can gain substantially during pregnancy but still lose condition because the combined weight of foetuses, membranes, uterus and uterine fluids will be approximately 15 kg for a 75 kg ewe carrying twins just before she is due to lamb. The target for lowland ewes should be a body score of about 3 at lambing and certainly not less than 2½ and about a half-score lower for hill ewes. Otherwise there is a danger of eroding the gains made through having ewes in optimal condition at mating.

ENERGY AND PROTEIN*

Apart from accessory food factors such as minerals and vitamins the value of a food is determined by its metabolisable energy and its digestible crude protein. The concentration of metabolisable energy (ME) which is defined as the difference between the gross energy of the ingested food and the energy of the excreta (dung, urine and methane gas) is expressed in terms of megajoules (MJ) per kg of dry matter (DM). Poor quality roughages such as over-mature or weathered hay have a value of about 7·5 MJ per kg of DM whereas cereals, at the other end of the scale, have values in excess of 13 MJ/kg DM. Values for the main feedingstuffs are given in Appendix III. Even when eaten to full appetite poor roughages have no more than a maintenance function with nothing over for production which includes liveweight gain, foetal growth and milk production.

Crude protein consists of true protein and non-protein sources of nitrogen. Over the past few years there have been substantial advances in knowledge of the fate of protein in ruminant nutrition. Only a varying proportion of the protein, the so-called undegradable dietary protein (UDP), goes from the rumen to the true stomach where it is broken down into constituent amino-acids before passing into the bloodstream. The balance which is known as rumen degradable protein (RDP) is broken down into ammonia by the micro-organisms in the rumen. A varying proportion of ammonia is used in the synthesis of microbial protein by micro-organisms and in this a principal factor controlling the level of recovery of ammonia is the amount of readily available energy in the rumen. Eventually the microbial protein passes through to the true stomach and small intestines where it, along with the UDP that passes through the rumen unchanged, is eventually digested. There are two important practical points arising from this. First, where there is a high proportion of RDP the dietary protein requirement is higher than that where the RDP component is relatively low. The toll that the rumen

* A more detailed account of ewe nutrition and the interactions of food components is provided by the Sheep Improvement Services of the Meat and Livestock Commission in its publication *Feeding the Ewe*.

micro-flora exacts from dietary protein has its compensations in that they are also capable of synthesising microbial protein from nitrogen-rich compounds like urea which is often a constituent of proprietary feeds including the feed blocks that have a growing popularity with hill farmers.

NUTRITION OF EWES BEFORE AND AFTER MATING

Immediately after ewes have been separated from their lambs it is sensible wherever this is feasible, to put them on short commons for about a week (and possibly longer in the case of ewes that have reared singles), in order to halt milk secretion which can be considerable in deep-milking ewes even after sixteen weeks of suckling by twin lambs. At this point there will usually be about twelve weeks before ewes are joined with the rams. The flock should then be graded for condition and any ewes with a score of $2\frac{1}{2}$ or higher can be used safely for about six weeks as animated mowing machines turning any rough growth that has accumulated over the season into dung and urine to create a fresh base for autumn growth that is stimulated by a late summer nitrogen dressing. Lean ewes with a rating of 2 or less will generally be among the best mothers in the flock and if they are to repeat their performance they will need the whole of the available three months on reasonably good grazing before being mated to achieve a target body score of $3\frac{1}{2}$. Supplementation of pasture with concentrates should be unnecessary except during a severe drought. Pasture is by far the cheapest nutrient and it must be exploited to the full because sheep, even highly fecund ewes, are not efficient food converters. It is better to reserve concentrates for a later stage in the reproductive cycle when pasture is in short supply and the physiological needs of the ewes are at a peak as they are in late pregnancy and early lactation.

A reasonably high plane of nutrition must be maintained over mating and for at least a month after. There are always fertilised ova that do not produce lambs and this loss can be aggravated by short periods of severe malnutrition in the period before the developing embryos are firmly attached to the uterus. For a lowland flock lambing in the second half of March this means good grazing until at least the last week of November. Upland and hill flocks which are mated to lamb somewhat later run the risk of nutritional upsets resulting from early snowstorms and this emphasises the importance of having reserves of good hay on hand to meet such a contingency. As was pointed out previously the middle third of pregnancy is a less critical period and any mature ewes in reasonably good condition can be given the job of cleaning up any rough growth remaining on pastures and volunteer growth on stubbles where these are available. The best policy is to 'mob stock' the ewes, that is to concentrate them for a short period on each field in a rotational sequence that can usually be repeated in an open winter. A ewe about 70–75 kg liveweight with a condition score of 3–3½ will need about 10 MJ of ME and about 140 g of crude protein per day for maintenance which is equivalent to an intake of 1.5 kg of average hay (8 MJ/kg DM). It is a matter of judgement whether the available grazing is sufficient to provide for this level of nutrient intake and if it is insufficient, which is the likelihood from about the second week of January if more than seven ewes are being wintered per hectare of grassland, and some pasture is being reserved for ewes after lambing, then it will be necessary to supplement available grazing with hay or silage. In fact on strong land with as many as ten ewes being wintered per hectare it is doubtful whether they should be on the land after the turn of the year. The point will be considered in greater detail in a subsequent chapter dealing with inwintering.

Two-tooths, ewes with a condition score less than 3, and any broken-mouth ewes will need rather better mid-pregnancy treatment which may come from first access to grazing on a leader–follower basis or from more liberal supplementation with good hay (9 MJ/kg DM). A 70 kg Mule ewe with an intake of 1·5 kg of hay of this quality plus some grazing would be getting sufficient energy and protein for maintenance and for an appreciable improvement in body score.

NUTRITION OF EWES BEFORE AND AFTER LAMBING

The nutrition of ewes over the 6–8 weeks immediately prior to lambing is of critical importance because it influences birth-weights, viability of lambs, milk yield and, in extreme cases of underfeeding, acceptance of lambs. Wallace and Hammond, working at Cambridge in the 1940s, were the first to provide experimental evidence of the importance of pre-natal nutrition and particularly the danger of drastic reductions in nutritive levels for ewes that have been well fed earlier in pregnancy. They showed, in effect, that low-level feeding over the first two-thirds of pregnancy followed by liberal feeding until lambing was just as effective as high-level feeding throughout pregnancy and it resulted in a more economical use of food. The reverse regime was catastrophic in terms of lamb survival, milk yields and health of ewes.

These Cambridge results, which have been substantiated and enlarged by other research workers under a range of environments, are understandable when one examines developments during pregnancy. In twin-bearing ewes the combined weight of foetuses, membranes, fluids and uterus at 90 days is just about one-third of the weight attained immediately prior to parturition while approximately 70 per cent of foetal growth takes place in the six weeks before full-term. There is also considerable mammary development including secretion of colostrum and all this requires additional nutriments for there is a limit to the extent to which a pregnant ewe can safely draw on body reserves.

In theory it would be advantageous to maintain a condition score of $3\frac{1}{2}$ in lowland ewes but because the last six weeks of pregnancy normally coincides with a period of limited availability of grazing and therefore a high dependence on conserved roughages it is not easy in practice to maintain condition without recourse to concentrate feeding, which is expensive. There is another important factor in this. As a ewe approaches lambing her appetite for bulk foods can become progressively smaller if the hay or silage is of low digestibility and has a low rate of passage through the digestive tract. A principal factor in a high incidence of pregnancy toxaemia (twin-lamb disease) is often a failure to notice a reduction in the intake of bulk food that has been offered *ad lib.* and to compensate for this by increased concentrate supplementation. A fall of about half a point in the condition rating over the last half of pregnancy will not generally have deleterious effects provided it is a gradual process and not due to an abrupt reduction in nutrient intake such as that occurring during snowstorms. Table 9 exemplifies levels of feeding that could be adopted in the later stages of pregnancy where the contribution from grazing is minimal.

Table 9. Levels of feeding for twin-bearing ewes in late pregnancy*

	kg/day	kg/day
Ewe weight	50 kg	70 kg
Six weeks before lambing		
Hay or	0·85	1·0
Silage	2·6	3·5
Plus concentrates	0·3	0·35
Four weeks before lambing		
Hay or	0·85	1·0
Silage	2·6	3·5
Plus concentrates	0·45	0·55
Two weeks before lambing		
Hay or	0·85	1·0
Silage	2·6	3·5
Plus concentrates	0·6	0·75

* Based on figures taken from MLC publication *Feeding the Ewe*

These rations would make sense only if a high-quality hay or silage were on offer, that is to say a hay that has been cut at the early flowering stage and cured with minimal weather damage or a silage cut at ear emergence and then wilted so that the final silage, following a lactic fermentation, has at least 25 per cent DM. Poor-quality silage can be a disaster in ewe nutrition, resulting in a loss of condition and broken fleeces, so silage intended for sheep should be conserved with as much care as that intended for high-producing dairy cows.

If the ensiled material has been treated with formic acid there will be an increase in the proportion of dietary protein that passes to the true stomach and this will permit economies in the protein component of the concentrate mix. Specimen concentrate mixes are given in Appendix IV. It is stressed that if lower quality hay or silage is on offer than that envisaged in Table 9 (10 and 11 MJ kg DM respectively) then a proportionately higher allowance of concentrates must be fed.

The above rations are designed for ewes carrying twins. It is recognised that they will be over-generous for ewes carrying singles and insufficient for ewes with triplets. This underlines the importance of using bodily scores as a means of giving low-condition ewes some extra feeding. Ewes with a condition score of 3 will be well capable of giving triplets a reasonable start in life but it would be a different story with ratings of $2\frac{1}{2}$ or less. At the risk of being a little old-fashioned there is much to be said for the inclusion of some roots, usually swedes, in the diet of ewes in late pregnancy and in early lactation especially if hay is the roughage. A succulent food in moderation has a mild laxative effect and this gives an edge to appetite.

FEEDING OF LACTATING EWES

Nutrient requirements are at a maximum at about 2–3 weeks after lambing for at this stage normal lambs are suckling repeatedly and are still entirely dependent on their mother's milk. It is not until the end of the fourth week that they eat appreciable amounts of solid food. Peak milk yields are reached by ewes with twins about three weeks after lambing but peak appetite is not reached for at least a further fortnight. The nutrient requirements of a ewe with twins during the first month are about 70 per cent higher than they are immediately prior to lambing. Even with plentiful spring grass, which stimulates appetite, and supplementary concentrate feeding a deep-milking ewe will almost have to draw on her body reserves to provide for her lambs' needs. This underlines the importance of ewes lambing in good condition for a plentiful supply of milk over the first six weeks is a key factor in producing a good crop of lambs, particularly where there is a high proportion of twins.

There are considerable differences between the requirements of ewes with singles and those with twins and triplets and it is sensible to separate these categories for it is wasteful to give all ewes the same level of concentrate feeding. Where there is an appreciable number with triplets there will be grounds for separating them from the ewes with twins and in addition to a rather more liberal concentrate allowance for the ewes creep feeding of the lambs would merit consideration because the milk flow of a ewe with triplets is only 10 per cent higher than that of ewes with twins.

It is not possible to give precise diets for a fat-lamb crop over the first six weeks of lactation because there is so much variation from farm to farm and from season to season, and with it a considerable need to exercise judgement. The onset of pasture growth will not only depend on season but also on management factors such as choice of species and fertiliser use. It is unlikely, however, that there will be sufficient grass to satisfy the nutritive requirements of the ewe with twins until well into April in the south of Britain and considerably later in the north. This is a time when a clamp of swedes or mangolds can be a valuable reserve as an alternative to high-quality silage if pasture is slow in coming away. It is not unreasonable to allow as much as 0·9 kg of concentrates daily for a 70–80 kg ewe with twin Down-cross lambs until the fourth week when they can be gradually reduced to zero at about the sixth week, again depending on the availability of grass. A maximum allowance for ewes with singles would be normally 0·5 kg of concentrates with feeding terminating after a month. Because a ewe in a negative nutritive balance is less able to draw on her body reserves for protein than she is for energy it is important that concentrates should have added protein, advisedly including a source such as fish or soya meal with a high proportion of undegradable protein if pasture is in short supply and hay or silage feeding has to be maintained well into lactation.

Contrary to common belief there is nothing to be gained from grinding cereals that are fed to sheep. In fact anything beyond rolling is a detriment. Apart from the cost there appears to be a greater risk of acidosis with finely ground meals when ewes are getting more than their fair share of offered concentrates. Sheep are able to digest whole grains and the apparently undigested grains that occasionally appear in the faeces are of no practical importance. Among other things there is less wastage with whole grains and a pelleted protein supplement when concentrates are fed out of doors in high winds.

NUTRITION OF HILL EWES

The basic principles in the nutrition of hill ewes are the same as those for lowland and upland ewes but naturally their application is considerably different. The hill sheep farmer's problem is again one of making the most economic use of the available grazing which is of much lower quality and less plentiful than that on lowland farms. On top of this his flock is subject to greater weather hazards and, of necessity, he must use breeds adapted to difficult environments. This in effect means that they are individually less productive than lowland sheep in that they are smaller, have a more primitive body form, and generally are less prolific. Inputs such as fertilisers and supplementary feeding must be on a much lower scale than those appropriate to lowland conditions and this necessitates more modest management targets. Even a condition score of 3 at mating, which should be the minimal target for lowland ewes, is out of the question for Blackface or Swaledale ewes. Apart from the cost, which would be prohibitive, there is the danger of too many lambs and all that this implies in respect of nutrition before and after lambing on hill farms with a limited area of improved pastures that have also to provide winter fodder. Hill ewes in good condition at mating are quite capable of producing as many twins as singles and such prolificacy could be embarrassing.

A condition rating of 2–2½ at mating is a realistic target and this must be obtained from grazing, following weaning in August to give at least twelve weeks before the tups are joined in November. Here, as workers in HFRO have shown, especially at Sourhope in the Cheviots, lies the importance of subdivision of the more productive parts of the hill, in addition to in-bye land, so herbage can be reserved for physiologically critical times such as the lead up to mating and the month immediately following, and the six weeks before lambing. During the second and third months of pregnancy hill grazing will often suffice if the land is clear of snow but it is an essential precaution to have strategically sited reserves for storm feeding. Nutrition of ewes on the hill can be greatly improved by block feeding from early in the New Year until March when it is advisable to introduce concentrates into the diet. The old idea that it is dangerous to supplement the diet of hill ewes because this would impair their hardiness is outdated nonsense and increasingly hill farmers are realising the importance of some extra feeding. Blocks, which sometimes seem to be over-priced for what they contain, can have an important part to play in this because of their convenience in feeding. Feeding of home-mix concentrates with ewes dispersed over the open hill is not very practicable and for this and other reasons concentrates are best reserved until nearer lambing when ewes are on enclosed land.

There are a variety of blocks available, often backed with sales talk that may give an exaggerated impression of their value. They are in no sense a complete food but they can have a valuable supplementary function for stock on low-quality grazing. This was established many years ago in southern parts of Africa with cattle on veldt grazing which has a very low nutritive value in the dry season. Access to licks containing urea with maize meal and molasses as a carrier and an immediate source of energy and common salt to keep urea intake within safe limits, increased microbiological activity in the rumen. This in turn speeded up digestion and the rate of passage through the gut and led to an increase in roughage intake of approximately 50 per cent as well as a substantial improvement in thrift. Many blocks on sale in Britain contain urea and they similarly promote a higher intake and a better utilisation of roughage resulting in an appreciable lift in the energy intake of hill ewes. From about a month before lambing and subsequently until ewes return to the hill it is generally advisable that concentrates should supplant blocks and that these should include a protein source such as fish meal which has a high proportion of undegradable protein. Levels of concentrate feeding have to be modest over this period as compared with lowland management starting at about 100 g per day at the beginning of the fifth month of pregnancy rising to 400 g at lambing, and tailing off to zero four weeks after lambing except in the case of ewes rearing twin lambs. Here the peak allowance could be as high as 500 g per day and supplementation might be maintained until the end of the fifth week after lambing if there is insufficient grazing. A great deal depends on a farmer's judgement. On the one hand he cannot afford to let lambs deteriorate because of insufficient milk but on the other hand he cannot be extravagant with expensive inputs in what must essentially be low-cost farming. An upper limit of about 20 kg of concentrates per ewe is a reasonable budget figure.

Chapter 5
FLOCK MANAGEMENT

GENERAL CONSIDERATIONS

Given the wide range of conditions under which sheep are kept it is not possible to describe management practices with a general application. In this respect sheep farming is very different from pig and poultry production where there is a considerable measure of environmental control and where the enterprises are, for most practical purposes, independent of other farm activities. At one extreme sheep are expected to give an economic return at the very limit of agricultural use of land, where they can be subjected to severe climatic stresses, while at the other extreme of good lowland farms they are usually integrated with other enterprises such as cash cropping or some branch of beef production. Nevertheless there are some basic aspects of management that have an application whether a flock is located on low ground or on high hills.

We have already considered one important management factor, namely nutrition, in this way and there are a number of others that affect the viability of a sheep enterprise no matter where it is located; for instance choice of breed, flock hygiene, systems of grazing and routine flock management. It is necessary, too, to examine them in a context of economic feasibility rather than one of biological possibility. For example, if hill ewes are mated in a very good condition there could be an embarrassingly high proportion with twins in the following spring relative to the available area of improved pastures, for with typical moorland grazing, it is usually preferable to have a ewe with one good lamb rather than two runts.

There is the reverse situation on a lowland farm with good pastures, for here the aim must be one of maximising the number of ewes rearing doubles. Though a single lamb may be sold somewhat earlier in the summer to catch the market before its seasonal downturn, a ewe, together with her twins taken to slaughter weight, requires less than 20 per cent more food than her counterpart with a single lamb to produce twice the amount of meat. The assumption in this calculation is that there is no protracted store period following weaning because this would increase the food intake of the twin lambs.

CHOICE OF BREEDING STOCK

With over forty distinct breeds, and Heaven only knows how many two- and three-way crosses to choose from, a sheep farmer, at least on paper, has an embarrassment of options; but in practice when the choice is narrowed down to what is appropriate for a given environment and a particular production objective the decision is usually straight-forward. For instance under hill-farming conditions there could be very little gained and a great deal lost from replacing local stock with another breed. Apart from the fact that an established breed has stood the test of time in a given situation it is most unwise to prejudice in any way the acclimatisation of a flock to a particular locality. If changes are to be made it is better to effect them gradually by the introduction of tups of the new breed. But in some regions where there is open grazing land and intermingling of flocks such an introduction could be regarded as an unfriendly act, especially on some of the Welsh hills where foreigners of any kind are not always welcome.

There are instances of the successful use of tups of another breed on existing hill stock, for example the use of the Swaledale on Blackface flocks on northern Pennine farms as well as the reverse mating and the North Country Cheviot on the South Country Cheviot on the hills that form the border between England and Scotland. Here an important factor contributing to the successful transition to a larger and more productive ewe has been the progressive improvement of pastures. In hill sheep farming as in racing it is very much a matter of horses for courses.

The upland or marginal-land sheep farmer has two main

choices for his ewe stock. Either he can purchase draft ewes from adjacent hill farms to mate with Longwool tups to cater for the generally lucrative trade in cross-bred ewes or he can choose a pure breed such as the Clun, North Country Cheviot or Derbyshire Gritstone. Choice will depend to some extent on traditional loyalties but with the growth in popularity of crosses, particularly the Welsh Halfbred, the alternative of a draft ewe flock has become more attractive. The choice of a pure-breed flock has the virtue that a farmer is able to breed his ewe replacements and also develop a more positive flock improvement programme than is possible when one is entirely dependent on stock bred by other farmers. An upland farmer with North Country Cheviots is in the fortunate position with this prolific breed of having the best of three worlds in that he can breed replacements from the top section of his flock, he can breed Scottish Halfbreds from the remainder, and he can continue to sell draft ewes before their reproductive potential has been completely exhausted.

The favoured ewe choice on lowland fat lamb farms, particularly north of the line from the Wash to the Mersey, must be one of the very efficient cross-breds by a Longwool tup out of a hill ewe. Not only are they very prolific but also they have pronounced mothering instincts and the liveliness of their lambs at birth as compared with lambs of the lowland breeds, which comes from their hill breed ancestry, is some safeguard against inclement weather at lambing. There is little to choose between the several cross-breds, and now that the merits of the Mule and the Greyface have been recognised there is no longer the large price differential between them and the more fashionable Scottish Halfbred. The general consensus among farmers with experience of these crosses is that the Mule and the Greyface are less expensive to winter than the Halfbred.

Welsh Halfbreds are smaller, less prolific and cheaper ewes than the north country cross-breds and yet they are remarkably competitive. During 1980 the Sheep Improvement Services of MLC conducted a survey of lowland spring lambing flocks, divided according to ewe type into three categories: small (typified by Welsh crosses), medium (typified by Mules), and large (typified by Scottish Halfbreds). Financial

Table 10. Lowland flock results by ewe type, 1980

	Small	Medium	Large
No. of flocks	64	222	62
Output/ewe (£)	32·3 (23·9)*	41·5 (26·7)*	44·2 (27·4)*
Flock replacement/ ewe (£)	3·6	6·6	7·6
Variable costs/ewe (£)	9·4	13·5	14·7
Gross margin/ewe (£)	19·3	21·4	21·9
Gross margin/lia (£)	246·6	247·4	242·7
Live lambs/ hundred ewes	133	153	159
Lambs reared/ hundred ewes	126	144	149
Stocking rate ewes/ha	12·8	11·6	11·1

* Figures in brackets are average sale prices/lamb

and performance data relating to these three divisions are given in Table 10.

On average the farmers contributing to the figures in Table 10 have reached a position where there is obviously no financial advantage for one type of ewe over either of the other two. The smaller lamb crop from the Welsh-type ewes and the lower price returned per lamb are largely compensated for by a lower cost of replacement and lower feed and forage costs, while the small difference in gross margin per ewe is made good by an appreciably higher stocking rate so that gross margin per hectare is practically the same for all three types.

Unfortunately there are no figures available to provide a comparison of pure-breds like the Romney and the Devon Longwool with cross-bred ewes under lowland conditions. It is suspected that the continued faith in these two breeds in their native parts owes a lot to local loyalties (or less charitably to local prejudices) because they are not fecund breeds and lamb sales per ewe are such an important factor in flock income. This criticism does not apply to the Lleyn which is outstanding among lowland ewe breeds for its prolificacy. A factor in maintaining the popularity of lowland ewes is that a farmer is able to breed his own replacements or buy them locally and this is important in areas that are at a distance from the sources of cross-bred ewes. It costs money to shift

Mule gimmers from Barnard Castle or Penrith to Kent, and often a farmer has to depend on an agent to make the purchase and arrange transport. It is not surprising if he prefers to go to Ashford market to buy the local article which will make good some of its deficiencies as a mother by heavier clips and lower depreciation.

FLOCK MAINTENANCE

Hill farmers are in the fortunate position that, with the exception of tups, they are able to breed their flock replacements. Their primary concern is the maintenance of an age-balanced flock with an annual output of draft ewes that will make a substantial contribution to farm income. Upland farmers with self-contained pure-bred flocks have the same objective, for instance with Clun flocks in the Welsh Marches they are able to sell annual drafts of breeding ewes that are still in their prime to Midland farmers who reap the benefits of having ewes at the most productive stage of their lives.

Most lowland fat lamb producers have to maintain their flocks by annual purchases and it matters that they do this as cheaply as possible because ewe depreciation has now become a major cost component of fat lamb production. It is a rather intimidating thought for a sheep farmer when he buys 100 Halfbred gimmers that in five years time they will possibly have reared 700 lambs and the eighty-five survivors (if he is an average farmer) will probably be worth only half the price per head that he paid for them.

There are several ways of reducing the impact of ewe depreciation. One that has been attempted with some success is the retention of double-cross home-bred ewe lambs such as Suffolk x Mules that would otherwise have gone for slaughter. This makes sense when the cost of cross-bred ewe lambs is substantially more than the price realised for slaughter lambs, for MLC records show that the mothering performance of these double-cross ewes by the Suffolk is only marginally behind that of the cross-bred ewes from which they have been bred. Suffolk-cross ewes have an additional virtue in that, like the pure Suffolk, they come into season rather earlier in the autumn than Longwool crosses with hill breeds and this is advantageous in early lamb production.

Gimmers have long been the most expensive of all breeding ewes and many farmers have switched to the purchase of ewe lambs with the hope in an inflationary situation that the differential plus the value of their hogget fleeces will lower their cost at point of entry to the breeding flock. They have gone even further by mating them so that they drop their first lambs as hoggets. This has been successful judging by the number of farmers now following this practice and it does not appear to have a deleterious effect on life-time performance of ewes (see Figure 1); in fact the general experience is that ewes lambing for the second time at two years of age are better mothers than those lambing for the first time.

MLC records over a six-year period covering a variety of crosses reveal that about three-fifths of ewe lambs that are put to the ram are mated successfully and, as one would expect, the success rate is higher where lambs are well grown. Where the average weight of the ewe lambs at mating exceeded 60

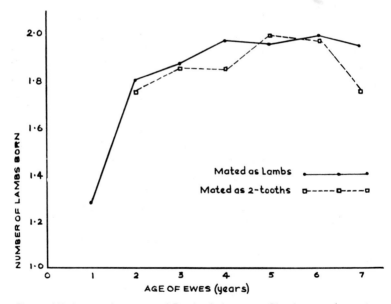

Fig. 1. Lifetime performance of Border Leicester x Cheviot ewes mated as lambs or yearlings.   From C. Yalcin

per cent of the normal adult weight 72 per cent of the hoggets produced lambs as compared with 49 per cent where ewe lambs mated at less than 60 per cent of the mature weight. There do not appear to be material differences in conception rates between the different cross-breds in the study but rather surprisingly because the Suffolk is a relatively early maturing breed its cross-bred hoggets were ten points below the figure of 71 per cent recorded for a composite group of Mules, Mashams and Greyfaces. Farmers with well-grown ewe lambs, that are run with experienced rams and with safeguards against checks immediately following mating, can fairly confidently budget on 80 per cent giving birth to an average of about 1·2 lambs. One should not, however, expect a yearling sheep to rear twin lambs because this will impair growth of both mother and offspring. Because there is usually a higher proportion of singles it is a misplaced kindness to have ewe hoggets in a high condition at lambing because this will increase the chance of difficult deliveries.

Reductions in the impact of depreciation can be made at the other end of the life span of breeding ewes. Many fat lamb producers sell their breeding ewes immediately they are seen to have broken mouths in the belief that they have reached the end of their useful lives as fat-lamb mothers. This is not necessarily the case as reference to Figure 2 will show. Even at nine and ten years old, provided they are in a fit condition at mating and they are not subjected to nutritional stresses as they approach lambing, they are capable of rearing satisfactory lambs. Plate 3 illustrates this point very convincingly. The fifty-five Clun ewes in the picture had had not less than six crops of lambs and yet they produced 102 lambs of which two were born dead, one was destroyed because of deformity and one was taken by a fox while the remaining ninety-eight lambs were well above average weights at weaning. Admittedly the ewes were given preferential treatment from mating. They were housed in mid-December, some two months before they were due to lamb, and they were trough-fed over that critical last third of pregnancy.

Two months prior to mating the salvage value of these mainly broken-mouthed ewes would have been less than one-sixth of the value of their lambs at weaning and the

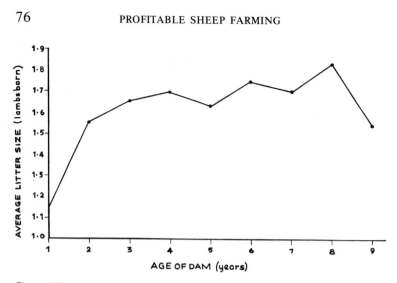

Fig. 2. Effect of age on size of lamb crop from Clun Forest ewes.

From M. Bichard

PLATE 3
Nucleus Clun flock at Cockle Park: 55 six- to eight-crop ewes with 98 lambs.

Photo Farmers Weekly

additional concentrates that were allowed them and their lambs did not amount to more than 25 kg above the normal for ewes lambing in February. Their salvage value after they weaned this crop of lambs differed very little from that twelve months previously. Provided ewes are sound in their udders and do not suffer from some chronic infirmity they are still capable of rearing satisfactory lambs if they are given the right treatment, which could include winter housing as well as extra feeding. The additional cost of this will be repaid by the low ewe depreciation costs that these extra crops of lambs will carry.

TIMING OF LAMBING

Leaving aside autumn–early winter lambing based on Dorset ewes, which is a highly specialised operation catering for a luxury market, the nature of fat lamb production is such that pasture as grazing, which is the cheapest source of nutrients, must be the principal food. With most production systems ewes and their lambs receive over 90 per cent of their nutrients from grazing between the onset of growth in the spring and its cessation in the autumn. Though the pattern of growth differs in response to climatic variations, on average under lowland conditions about 50 per cent of grass growth comes between mid-March and mid-June, about one-quarter over the next two months and the remainder is produced between mid-August and early November.

The balance of production as well as availability of grazing can be manipulated to some extent by the sowing of early- and late-growing species and the application of nitrogenous fertilisers while some of the autumn flush can be reserved for November-December grazing provided the locality is reasonably free from early frosts. There is a limit to the effective growth manipulation and so the principal effort must go into securing the best coincidence between flock appetite and the availability of grazing. There are, however, other considerations that come into this planning. In a study undertaken by MLC over a five-year period, lambs from flocks lambing in January–February had a price advantage of nearly 12 per cent over lambs from spring lambing flocks (March–April). A farmer has to weigh the cost of extra hand feeding in early-lambing flocks and the probable reduction in the numbers of lambs reared per ewe against this price advantage which may not hold in all parts of the country.

Possibly the best compromise is to commence lambing about a month in advance of active pasture growth. Inevitably, despite one's best efforts, there will be a spread of at least a month between the start and finish of lambing and it is important to have the majority of lambs eating appreciable quantities of grass when growth starts to accelerate in late April–early May. Another consideration is a fair proportion of lambs in a marketable condition before prices start a severe downturn and also to reduce the level of stocking as grass growth begins to fail for it can be a source of embarrassment to have a lot of store lambs on a farm in September when one should be giving ewes better treatment. If the farmer is in the fortunate position of having ample in-wintering accommodation he may be tempted to start lambing with a portion of his flock, preferably his oldest ewes, even earlier than a month before pasture starts to grow and this could make good sense provided there is really good hay and/or silage and possibly mangolds or swedes available to keep up the flow of milk until there is the further stimulus of fresh grass. Among other things old ewes, destined for slaughter after lambing, will come on to the market when prices for this class of sheep are still reasonably firm.

Under typical hill conditions it is unwise to start lambing until the end of April and even then it is advisable to lamb on in-bye fields where, hopefully, there will be some freshening of the pastures. Little in the way of growth can be expected on the open hill, except in favourable west coast situations, until well into May.

MATING

It was stressed in the chapter on nutrition that it is essential to have ewes in a fit condition in order to maximise the lamb crop. This applies to rams as well because a tup will not do his job properly if he is lame or if he is not literally in the pink of condition for it is only then that his *libido* will be at a maximum. Normally a ratio of one ram to 50 ewes is recommended but a fit ram can handle many more especially where a ram and his mates are in separate enclosures. Many farmers are happy with one ram to 80–100 ewes with paddock mating in the knowledge that a ram with a small mating group will often indulge in repeated mating but where there is a ewe pressure he will spread his favours to equally good effect. The above ratios apply to shearling or older rams; advisedly they should be halved with ram lambs. A wider ratio than 1–50 is not recommended under hill conditions and there it is important to round up the ewes at least once- but preferably twice-daily in order that all rams and ewes are in frequent contact. Left alone a ram can be content with a small harem and there can be an appreciable number of unmated ewes.

It is important to have as compact a lambing as possible. There are several ways of effecting this. Fit ewes come to the ram more readily than thin ewes and this is another bonus from mating ewes with a reasonably high condition score. Crossbred ewes like the Mule or the Halfbred normally come to their sexual peak in mid-autumn so if one attempts to mate them in late August-early September conception is usually a drawn-out affair. The use of hormone-impregnated sponges is one way of getting these ewes into a receptive condition for early mating but this is fairly expensive and possibly the best method is that of running vasectomised rams with the flock for about 16–18 days before it is joined by entire rams. Vasectomised rams, commonly known as teasers, retain their sexual drive but are unable to mate effectively. Vasectomy is a simple operation for a veterinary surgeon but it is not reversible so it is not one for rams with a further breeding requirement. It is unwise to exceed 16–18 days for the stimulus seems to disappear with a longer association. Farmers who have adopted this technique for the first time have been pleasantly surprised by its efficacy with as many as 90 per cent of the ewes lambing within a fortnight. This is highly advantageous because it is difficult to maintain high-intensity shepherding over a protracted lambing if a shepherd has other responsibilities.

Paddock mating, that is splitting up the flock into convenient mating groups each with one ram, is preferable to an arrangement where several rams are turned out with a large flock for then it is impossible to tell whether every ram is doing his job properly. With paddock mating and brisket raddling of the rams, with a change of colour every 14–15 days it is possible to check returns to service and so pick up any sterile rams with minimal delay. The best succession of colours is yellow, red and blue. Raddling is not only useful for checking conceptions but also it helps in sorting ewes into groups based on expected times of lambing. Apart from the convenience of this at lambing it is possible to make considerable economies in the supplementary feeding of concentrates. One does not usually start this until about six weeks before the due date and it is wasteful to feed as though all the ewes were due to lamb 147 days after the first mating. When the ewe lambs are being mated it is advisable to use an experienced ram. Often a young ram will mount and mark its mate without effecting a proper connection and as a result there will be a disappointing conception rate. Ram lambs are particularly bad in this respect.

LAMBING

Knowledge of how to handle a flock at lambing cannot be learned from a book but only from experience. Fortunately if the ewes are in good condition and they have not been subject to any upsets, nutritional or otherwise, there will be few complications and most ewes will need supervision rather than assistance. If a ewe needs help to correct an abnormal presentation then attention must be paid to the simple elements of hygiene for otherwise there is a danger of septicaemia.

But the main causes of loss of lambs at birth are not due to lambing difficulties but to mismothering and perishing as a consequence of exposure. In one MLC study 41 per cent of the recorded deaths, soon after birth, were due to mismothering and starvation and it is probable that a proportion of the 20 per cent where no cause was determined also came from this cause. Leaving aside lambs dead at birth, about two-thirds of the subsequent mortalities occur in the lambs' first week. Because this is the stage when they are most vulnerable it is also time when the greatest dividends will accrue from close shepherding.

A large crop of lambs is vitally important in the economics of fat-lamb production and the more intensive the system becomes the greater is the need for continuous supervision of lambing, at least during its peak, and also for the provision of artificial shelter. This in its simplest form need be no more than a scatter of bales in the lambing field or, more elaborately, banks of bales or thatched hurdles or fences.

None of these devices compares with indoor lambing and one of the side benefits of inwintering a flock is that this accommodation is also available for lambing so that risks of exposure are reduced to a minimum. Essential extras with indoor lambing are infra-red lamps, such as those used for newly born pigs, and individual lambing pens about $1 \cdot 5$ m² in area. As soon as a ewe has lambed she and her lambs can be penned and thereafter they are under complete control. One can then make certain that the lambs are suckling, for possibly the best insurance for survival is mother's milk. A valuable aid with a weakly lamb that does not suckle properly is the insertion of a flexible rubber tube down the gullet to ensure that hand-milked colostrum reaches the stomach. A high proportion of lambs that perish outside are the first born of twins and triplets which wander away from their mother without having suckled properly.

Normally a ewe and her lambs need not remain in a lambing pen more than a day. They can be bulked in larger pens or if the weather is good they can be turned out to a sheltered pasture but with this proviso, if rubber rings are used for docking and castration when lambs are only a day old, do not turn them out before they have fully recovered from the application. There is a danger, especially with male lambs, of separation from their mothers and if the weather is bad chance of their survival is greatly reduced. In fact it is preferable to delay application until lambs are at least a week old unless one is prepared to ensure that they are really on their feet and following their mothers before they leave the fold.

Alternative methods of docking and castration are with a

knife or a searing iron and bloodless castration can also be effected by using special pincers to crush the cords. Numerous tests of the several methods on growth have established very small differences between them but one definite finding is that the earlier the operation is completed the less will be the check on growth. Castration and tailing with knife and searing iron are best undertaken under dry conditions using a temporary pen put up specially for the purpose on clean land because castration using permanent yards greatly increases the risk of infections leading to joint-ill. Needless to say disinfectants should be used as a further safeguard. There is also a risk of joint-ill from lambing on the same field year after year or in yards. The simplest method of control is to treat the navels of new-born lambs with a mild disinfectant or surgical iodine.

A quota of triplets, and sometimes quadruplets, is inevitable with well-fed, highly prolific ewes and it is important to make the most of them and also of any orphan lambs. It could be worthwhile with a large flock to have some form of artificial feeder, where surplus lambs have access to reconstituted milk, later supplemented with dry food; but most shepherds will attempt to set spare lambs on ewes that have either lost their lambs or are rearing singles. One reasonably effective trick with ewes that have lost their lambs is to cover the foster lamb with the skin of the dead lamb.

The task of mothering-up is much harder when the foster mother still has one of her own lambs. Possibly the easiest method is to have a mothering crate where a ewe can be firmly secured by the head so that the foundlings are able to suckle without molestation. It is said that a ewe will accept a foster lamb once her milk has passed through the lamb's body but this is possibly no more than an estimate of the time required for acceptance. There is also the possibility of allowing ewes to retain their triplets, especially where they can be segregated on good grazing and receive a concentrate supplement. It might be advisable, however, to wean the smallest lambs of the trios at 6–7 weeks and put them in a feed-lot on hay and concentrate diet. Lambs make a remarkably efficient use of concentrates from this stage up to 15–16 weeks when they should be fit for slaughter.

SYSTEMS OF GRAZING

Much more important than the method of grazing, whether it embraces mixed stocking, set, rotational or creep grazing is the quality and cleanliness of pastures. Sheep, especially lambs, do not thrive on coarse, stemmy herbage. There is a saying on the Romney Marsh that anyone dropping a sixpence while walking across one of the renowned fattening pastures should find it if he retraces his footsteps. Another Marsh saying is that a fattening sheep should graze by day the blades of grass that have grown the night before. Possibly this is over-ambitious but there is no denying the value of a dense leafy pasture in promoting thrift in sheep, especially ewes and their suckling lambs.

But qualitative and quantitative sufficiency is not enough. Pasture must also be clean and in the context of fat-lamb production this, primarily, means freedom from parasitic worms, especially the round worms of the alimentary tract. The harm caused by these worms has been recognised for a very long time and assorted pieces of advice on how to minimise the problem have been passed on by one generation of shepherds to the next; for instance sheep should not hear the church bells on successive Sundays in the same field and a sheep's worst enemy is another sheep. There have also been traditional remedies which possibly did more harm than good, for instance nicotine and copper sulphate which could be as lethal for lambs as it was for worms.

Fortunately the position has changed dramatically over the past twenty-five years with the development of several highly efficient drugs and a growing knowledge of the biology of the principal round worms. It is now possible, by combining a dosing programme with the resting of pastures, to create conditions where the challenge from parasites is so small that they no longer noticeably affect thrift. These points are discussed in detail in Chapter 13 so at this point it suffices to say that hygiene, combined with strategic drenching with highly efficient drugs, unquestionably constitutes the greatest single advance that has been made in sheep husbandry in Britain during this century. Not only has it made possible increased intensities of stocking but also it has greatly improved the condition of lambs to the point where intensive fat-lamb production now has a stronger competitive position in lowland farming than it has had since the days when folded sheep were a cornerstone of cereal production.

Turning to systems of grazing, the best individual lambs will be obtained with set grazing (or non-rotational grazing), provided the stocking intensity is not so high as to create excessive competition for the available herbage. This was the conclusion reached at the end of the farm-scale comparison over several years of set and rotational grazing of ewes and lambs with two stocking intensities at Ruakura Research Station in New Zealand. The best lambs at weaning in most seasons were those from the medium intensity set grazing treatment while the poorest lambs came from the high intensity set grazing treatments. Lambs from the high intensity rotational system were intermediate in quality.

The practical conclusion at the end of this comparison was essentially this: if a farmer is reasonably content with a medium level of stocking there is no point in investing in close subdivision of fields but if his aim is one of maximising output then rotational grazing has considerable advantages and can justify the additional investment in stock and fencing.

The Ruakura study was on a year-long basis so treatment effects covered much more than the suckling period. Rotational grazing gave greater flexibility over the year in providing for the needs of the stock which were entirely dependent on grazing. We have a very different situation in Britain where supplementary feeding, apart from its contribution during the winter, is able to take the pressure off pasture in the early spring. The problem comes later in May and June when there is a big upsurge in growth and the ewes and lambs, despite their collective increase in appetite, are unable to control pastures. At this point closer subdivision and rotational grazing are advantageous because some pasture can be taken out of the grazing sequence for a light conservation crop. Most farmers, however, would be more inclined in these circumstances to draft in some store cattle to take the top off pastures when they are getting away from the sheep. In the opposite situation where there is a danger of insufficient grass to provide for the needs of the flock the British farmer has another aid denied his New Zealand counterpart on economic grounds, and that is the judicious use of nitrogenous fertilisers to boost growth.

A shortcoming of rotational grazing is that stock tend to become restless before a change of grazing. On the one hand there is the need to effect complete utilisation and ensure fresh regrowth and on the other there must be adequate nutrition for the ewes and their lambs. Even with the best compromise between these two objectives the flock tends to be on a saw-tooth plane of nutrition and this is reflected in its grazing behaviour. Sheep spend more time looking for food on their last day in a paddock than they spend on the first day.

It was primarily this observation that led to the development of rotational creep grazing at Cockle Park in 1955–6. Another factor came from a contemporary study of milk yield of ewes rearing twin lambs (see Figure 3). There is a highly significant correlation between milk yield and the growth rate of lambs between birth and eight weeks but from then on to

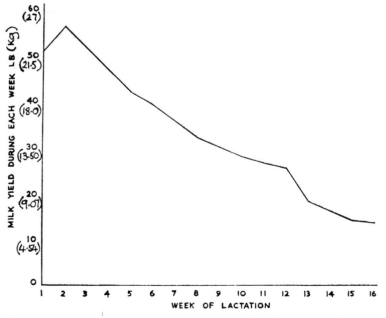

Fig. 3. Lactation curve of Border Leicester x Scottish Blackface ewes.
From A. Labban

weaning at sixteen weeks there is a poor correlation which is not surprising because of the lambs' increasing appetite for solid food. It was argued, why not give lambs first access to fresh pasture in an arrangement similar to that adopted for folded sheep on forage crops where the ewes were required to clear up after their lambs? This would minimise competition between ewes and their lambs especially over the later stages of suckling. The thesis was tested out experimentally in a direct comparison with normal rotational grazing and a 5 kg advantage at weaning was recorded for the lambs having prior access to grazing in the six-paddock rotation.

The system was then tested on a commercial scale at the University's Nafferton Farm in 1958 and subsequently for about fifteen years with considerable success. Scottish Halfbred ewes and their lambs, stocked at a rate of 20-22 ewes and 30-35 lambs per hectare regularly produced a net liveweight gain of at least 900 kg/ha between April and July on clean ryegrass and white clover pastures. The demonstration created considerable national interest and there were many attempts to adopt the system but these gradually petered out and now there are few farmers that practise this form of grazing despite the fact that there is possibly no system that compares with it in respect of efficiency of pasture utilisation.

But there are other considerations from a farmer's viewpoint than efficiency of utilisation. First it is a rather complicated system and it has to be made to work by attention to essential management details. Secondly, temporary netting fences for field subdivision are expensive in terms of materials and labour. Thirdly, the quality of lambs at weaning does not compare with that of lambs reared at a lower intensity of stocking with set grazing. It is better suited to the production of store lambs for further feeding rather than selling lambs fat off their mothers. Finally it is unsuited to small ewes, or even larger ewes after shearing, because they can be just as adept as their offspring in squeezing through the creep gates to the greener pastures beyond. In particular it is unsuitable where farmers are breeding their own replacements because creeping habits acquired as lambs remain with them in their adult lives.

WEANING

If ewes are taken away from their lambs rather then *vice versa* lambs will settle more quickly in their accustomed surroundings than they will on a fresh field. The ewes should be concentrated on a bare pasture for a few days, in order to dry them off as quickly as possible but within about ten days they should be sorted into groups according to condition as a first step in the programme of fitting them for mating.

Weaning normally takes place when the lambs are about 16 weeks though some farmers, with low stocking rates, delay weaning in order to draft as high a proportion as possible off their mothers. There is, however, little point in keeping lambs on their mothers after sixteen weeks under intensive conditions for by this time the average ewe will be giving very little milk and will be competing with the lambs for grazing. There is, in fact, good ground for appreciably earlier weaning in dry seasons when grass is in short supply because the double processing of grass into milk and then the milk into meat is biologically inefficient at the tail-end of a lactation.

As soon as the lambs are settled they should be dosed for worms and then moved on to clean grazing such as a conservation aftermath, preferably one with an appreciable white clover component, or to a forage crop such as rape or stubble turnips. An over-riding consideration in their management is the avoidance of any action which will prejudice the availability of clean grazing for the following season's lamb crop.

Chapter 6

INWINTERING OF EWES

Housing of sheep is a practice of long standing not only in countries with severe winters like Iceland and Finland but also in temperate countries. In many European countries sheep graze by day to return to folds by night but often this is a security measure to combat predators, particularly stray dogs. Inwintering of ewe hoggs was once common on hill farms in Scotland until the practice developed of sending them to lowland farms for the severest part of the winter but increased costs of off-farm wintering, combined with a shortage of farmers who are prepared to take sheep 'on tack', has reawakened an interest in home-wintering with some structural shelter.

The main interest today, however, is the housing of breeding ewes on lowland farms, especially those with strong land, where fat lamb production is being intensified. Twenty years ago inwintering of ewes was a novelty in Britain and was regarded by some as a nine-day wonder but now it is being gradually accepted as a standard practice, especially by farmers who have to be their own shepherds, which is increasingly the situation with the cost squeeze that farming is constantly facing. Not the least of the benefits of inwintering is the improvement in labour productivity and a lessening of discomfort of tending sheep in the worst weather of the year.

In theory a hectare of well-managed lowland pasture, with some additional help from concentrates (50–60 kg per ewe), will cater for the nutritional needs of 12–14 Mule or Halfbred ewes over the whole year as well as their 20–24 Down-cross lambs, up to an average age of about 150 days. In practice this target cannot be reached on heavy land if it is severely poached in the winter months. Normally from one-third to a half of the total annual production of pasture comes before the end of May but with puddling and poaching this proportion can be greatly reduced. Often there will be some compensating growth in mid-summer because of a stronger growth of white clover on pastures that have been punished in the previous spring but overall there will be an appreciable loss of production.

TRIALS ON NEWCASTLE UNIVERSITY FARMS

This point was investigated in the winter of 1964–5 at Cockle Park on a dense ryegrass and clover ley in its fifth year. Ewes were stocked at an intensity of 10/ha for the first half of the winter and 20/ha for the latter part of the winter to simulate farm conditions where half the grassland is rested from the New Year so there is some fresh grazing for ewes after lambing. Despite the winter being very dry with considerably less than the usual amount of poaching grass production was reduced by 20 per cent up to the end of June as compared with paddocks that had been fully grazed in the back end but had been rested from the turn of the year.

With a more typical Northumberland winter the loss of production would have been much greater. But it is not just the amount of loss that is serious but also its timing because April is normally the hungriest month of the year on a livestock farm when winter food reserves are often on the point of exhaustion while stock appetite is increasing rapidly. On top of this fresh grazing is vitally important in stimulating and maintaining milk flow which is so critically important in giving lambs a good start in life.

One can reduce the overall effects of poaching on a tillage farm by concentrating the flock on sacrifice fields ear-marked for late winter ploughing and a spring-sown crop but in most winters with heavy land such fields become quagmires by the middle of February and a farmer is usually forced to move the flock to pastures that are required subsequently for grazing or conservation. The damage will not be serious if the farm is blessed with some dry land or some old permanent pasture but the latter is now a luxury that one can afford only in small quantities on land well suited to cereal growing. Indeed with the substantially higher returns from autumn-sown barley and wheat as compared with spring-sown varieties and another valuable autumn-sown crop, oil-seed rape, sacrifice fields are also becoming something that farmers can no longer afford.

Our first attempts at inwintering breeding ewes at Cockle Park in 1960, in an endeavour to resolve some of the problems involved in high intensity sheep stocking on boulder clay soils, were modest and tentative. A hundred old Mules, with scarcely an incisor tooth between them, were wintered in a Dutch barn with an open yard on the south side and sheeting to ground level on the remaining three sides. They settled quickly into their new environment and, as a farmer from Kent remarked on one particularly miserable February day, all they required was television to have full home comforts.

Their basic diet was silage and hay to appetite until the last six weeks of pregnancy when they received concentrates which were gradually increased to 300 g daily. This was not enough because two ewes died from pregnancy toxaemia about a fortnight before lambing. At this point we had not appreciated the extent to which a twin-bearing ewe loses its appetite for bulk food, especially silage, as pregnancy advances. Possibly this was aggravated by confinement and lack of exercise but there was no more trouble after trough feeding was raised to 500 g daily.

These old ewes ended with just on a lamb and three-quarters apiece and this should have been better but at that time we did not realise the importance of more regular supervision of lambing when ewes are closely confined. These were growing pains born of inexperience but apart from them there were no major snags and we were given the confidence to inwinter a combined total of over 1,200 ewes on the two University farms, Nafferton and Cockle Park. There were Jeremiahs that forecast catastrophes, for instance a high incidence of pregnancy toxaemia, but this did not happen. In the winter of 1964–5 no ewe was lost on either farm before the commencement of lambing and the losses that occurred during lambing were less than one-half per cent—a quite remarkable result with the high proportion of ewes in the two flocks that had had more than five crops of lambs.

There was one occasion, however, when there were sudden losses in housed sheep at Nafferton due to pneumonia but this was with Swaledale lambs that were being autumn fattened on hay and cereals. The sheep shed at Nafferton was then an open-fronted building with a low pitched roof and a depth of about 15 m. The only provision for ventilation was the open front and an arrangement where the cladding on the back wall stopped short of ground level and air was able to move through this gap and over a low baffle wall that gave the sheep protection from draughts. That was the intention but, inadvertently, hay stored against the back wall prevented the ingress of fresh air and this resulted in humid conditions in the shed. As soon as the hay was moved the trouble ceased but as a further precaution ridge ventilation was subsequently installed.

The importance of good ventilation, combined with freedom from draughts, cannot be overstressed. Sheep are not happy in a stuffy humid atmosphere and if there is evidence of severe condensation in a building then one can be certain that it will be unsuitable for inwintering sheep because of the likelihood of respiratory troubles. Inadequate ventilation is probably the underlying factor responsible for most of the failures in housing sheep.

BENEFITS OF INWINTERING

There are now many sheep farmers who practise inwintering and the fact that their numbers are increasing is a tribute to its effectiveness. The principal advantages claimed for inwintering, provided general management matches the intention, are as follows:

1. There can be an increase in the area sown to the higher yielding winter varieties of barley and wheat because there is no longer a need for sacrifice fields. Apart from better yields there is less pressure on spring sowing and therefore a better prospect of the reduced area being sown when soil conditions are favourable. Additionally, with winter barley, harvesting can be started up to six weeks earlier in the summer, as compared with spring barley. This is of great importance in

late districts because the harvest day can become depressingly short in the tail-end of September when heavy dews conspire with reduced temperatures to push up drying costs.

2. There is a reduction in labour, and shepherding can become a part-time job even with a flock of 500 ewes, except at lambing when additional help is essential, provided that with foresight bulk food such as hay and silage is stored conveniently close to the inwintered sheep. With outwintering, once pasture no longer suffices, the shepherd, and sometimes a tractor driver as well, have to spend additional time carting food to their charges (see Plate 4). Much of the damage to pastures that results from outwintering comes from tractor and trailer wheels cutting into the soft ground to leave scars that sometimes remain until the field is ploughed.

3. The sheep, and the shepherd as well, have more comfortable lives with protection from the elements. This is of particular importance at lambing as newly dropped lambs are very vulnerable under stormy conditions.

4. Housing makes it possible to keep ewes longer without detriment to the lamb crop. With breeding ewes as expensive

PLATE 4
Outwintering can be expensive in terms of both labour and tractor use.

Photo B. Tyrer

as they have now become it is essential that flock depreciation is minimised for it is one of the major costs in lamb production. Housing not only protects these older ewes from climatic stresses but, with a division of accommodation into pens, it is possible to segregate sheep requiring preferential feeding from those that need no more than the standard flock management. Generally inwintering facilitates feeding according to condition score and, when ewes have been colour-marked at mating, according to expected time of lambing.

5. Unquestionably the most important advantage is one that has already been discussed at some length, namely pastures can be free of stock for at least three months from the turn of the year and not only is mechanical damage to soil avoided but leaf left on pastures is able to express its other function of growing more leaf so that there is improved grazing for the flock immediately after lambing.

Following this point we initially made the mistake at Cockle Park, which is an exposed farm, of lambing in early March and we found in late springs that the benefits of winter resting were lost by stocking too soon. When we switched to late March–early April lambing there was an appreciable improvement in six-week weights which are a good indicator of milk yields of ewes.

6. Inwintering can be helpful in the control of internal parasites. Apart from the pastures being free of sheep for at least a quarter of the year, ewes can be drenched while they are still inside against the so-called 'spring rise' in the output of worm eggs. When they return to pasture the risk of their reinfecting pasture is greatly reduced.

7. Finally there is the great boon at lambing of having the convenience of electric light when supervising the flock at intervals through the night. Every sheep can be clearly seen in a matter of minutes and ewes that have lambed can be eased into lambing pens without much bother. Contrast this with the nocturnal searching of a lambing field in dirty weather, torch or hurricane lamp in hand, to discover ewes that have dropped their lambs and are most unwilling to leave the spot where they have chosen to lamb for the relative security of improvised lambing pens!

DRAWBACKS AND THEIR ALLEVIATION

1. The principal criticism of the system is the considerable capital cost if new buildings have to be erected. Sometimes costs can be reduced by incorporating existing buildings in the layout but this can occasionally be at the expense of convenience in feeding and mucking out. A farmer going out of feeding cattle may have a wide-span building available and this can be ideal if it is possible to install demountable pens on either side of a feeding passage but it would not be advisable to put up such a building just to accommodate sheep for 15-16 weeks each year. The investment would only be justified with other uses, for instance covered shearing accommodation, implement storage, barn hay-drying, and floor storage of grain intended for disposal before the year's end. The great virtue of wide-span buildings is their flexibility but generally something more modest will be deemed advisable. Plastic-covered structures of the sort used by nurserymen have been used successfully and they have a reasonable life if they are well built and are located on a sheltered site.

An alternative is a simple lean-to which can sometimes be attached to an existing building, preferably so that it has a south-facing front. There is no need for all the accommodation to be covered. A suitable arrangement is one where a lean-to, about 5 m in depth, is fronted by a concrete yard of similar proportions. An allowance of $1 \cdot 5$ m² per ewe is ample. Only the covered portion will require straw bedding which is usually a much cheaper proposition than moveable slats which must rest on pillars to allow for a build-up of the dung that passes through the slats. About 3 tonnes of wheat straw is sufficient for 100 ewes over the course of a winter. The lean-to portion of this accommodation with a wire-netting front can provide suitable accommodation for rearing turkeys for the Christmas trade.

2. A second general criticism is that inwintered ewes require more roughage and concentrates to compensate for grazing. This is not a valid criticism when stocking intensities are so high that there is virtually no grazing after the turn of the year nor does it hold in those winters when there is snow cover on the ground from January to March.

On average, however, inwintered sheep require more hand feeding than outwintered sheep. In an MLC investigation where seventy inwintered flocks were compared with seventy comparable outwintered flocks the former received an average of 47 kg of hay equivalent and 51 kg of concentrates. The corresponding amounts for the outwintered sheep were 31 kg of hay and 44 kg of concentrates. The difference between the two treatments is not large and it could be a small price to pay for other benefits such as improved spring grazing, better labour utilisation or increased returns for cereals. Inwintering can only be properly evaluated in terms of its contribution to the management of a farm as a whole.

3. With close confinement in accommodation that is used by sheep every year there could be increased risks of disease. For instance an infection like orf can spread very quickly through a housed flock. This is a valid criticism of the system and it underlines the importance of prophylaxis and hygiene. Possibly the most serious risk is that of pneumonia but this seems to be a problem only if there is poor ventilation, accompanied by excessive condensation and wet litter. Lameness need be no more of a problem with inwintered sheep than it is with outwintered sheep, provided feet are kept trimmed and they pass regularly through a formalin bath.

4. The value of the fleece may be depreciated. This is a valid criticism but not a serious one if the sheep are properly fed. Severe checks, such as that from feeding poor silage, will cause breaks in the wool and sometimes lead to the loss of the whole fleece but this can also occur with outwintered sheep. Where sheep are feeding through a barrier, particularly one with sharp edges, there can be some loss of neck wool but this is usually very small. Soiling of the fleece through lying on faeces is of little consequence because stains quickly disappear when sheep return to pasture.

5. There can be higher lamb mortality if hygienic standards are low and if insufficient protection is given to newly born lambs. There is no evidence of higher lamb losses with inwintering in the MLC investigation referred to above but it cannot be denied that risks increase with the intensification of any branch of animal production, especially with young ani-

mals that have not yet developed antibodies. The answer to this, however, is the adoption of appropriate precautionary measures. For instance it should be standard practice to dress the navels of all newly born lambs with surgical iodine or a disinfectant to prevent the development of joint ill. As soon as a ewe has dropped her lambs she should be placed in a lambing pen and care should be taken to ensure that her lambs are suckled properly. They should remain in the lambing pen until it is certain that all is well with the family group, and as this usually requires a full twenty-four hours it is important to have a sufficiency of lambing pens.

GENERAL MANAGEMENT CONSIDERATIONS

If ewes are to be rationed according to body condition, age or expected time of lambing accommodation must be divided in such a way as to cater for these separate groups. The number per pen can vary quite widely: 20-30 ewes is a convenient number but fifty or even more is acceptable provided there is sufficient trough space so that all ewes get a fair share of the concentrates. An arrangement where concentrates can be placed in the trough without having sheep milling around has a lot in its favour. This can be effected by having a separate yard for concentrate feeding with batches being fed in succession from the same troughs.

Bulk foods should be on offer all the time so that sheep can feed to appetite. There is little to choose between hay and silage provided they are both of good quality and the choice is largely one of convenience based on general management considerations. There is something to be said for replenishing bulk foods in the morning because the fresh offering stimulates appetite and with the edge taken off their hunger ewes are a little less lively when it comes to feeding time for concentrates. A useful device for minimising wastage of hay is the Swedish hay box (see Plate 5) where a weld-mesh cover allows the sheep unrestricted access to hay but prevents them from pulling it from the box and dropping it on the ground.

It is strongly advised that lambs are not turned out to pasture before they are completely recovered from the effects

PLATE 5
Inwintering on straw; note the baffle walls and Swedish hay boxes.

Photo Farmers Weekly

of applying rubber rings which are particularly severe with male lambs. We realised the importance of this when we found that most of the mismothering that occurred immediately after turning out involved male lambs that were more sluggish and less inclined than ewe lambs to keep up with their mothers.

It became standard practice after this to bunch ewes and lambs as they came out of the individual lambing pens and keep them inside for a further 4–5 days and sometimes longer in bad weather before turning them out to pasture. Fortunately lambing coincided with the emptying of cattle yards as stores went out to grass and these gave the necessary elbow room. We found this extended protection was important in another respect. Weakly lambs, that would probably have perished with outside lambing, are given the chance to survive with inwintering but it is important to remember that they can still be very vulnerable for several days after their birth. There is little point in merely postponing their deaths for a couple of days by turning them out too soon.

Finally it must be reiterated that inwintering must be assessed not as an isolate but as an integrate of a complete farming system. Unquestionably it contributes significantly to the goal of making the most of the grassland devoted to sheep and unlike many innovations it does not complicate the winter management of sheep. If anything it simplifies this management in that weather hazards are minimised and a shepherd has a much fuller control of his sheep than he has with outwintering. The great pity is that hill farming is not more profitable so that farmers there could afford to invest in the necessary buildings and equipment so that they and their flocks are no longer vulnerable as they are at present with severe winter conditions that can decimate a flock.

Chapter 7

STORE LAMB FEEDING

ALTERNATIVE OF REFRIGERATED STORAGE

A relatively small proportion of Britain's lamb crop is sold for slaughter at point of weaning and there are several reasons for this. There is a demand for lamb throughout the year and though it tends to be stronger in the summer period, with most lambs being spring-born, heavy slaughterings at 14–20 weeks of age could swamp the market as well as lead to shortages at other times of the year. There used to be occasions in the late summer when producers have been their own worst competitors with frozen lamb from New Zealand making higher prices in the late summer on the Smithfield market than the fresh home-produced article.

Looked at purely from the viewpoint of efficiency of food utilisation it does not make sense for fat lambs to lose bloom just after weaning and then to take several months before they are again in slaughter condition, albeit at somewhat heavier weights. It is a wasteful use of food that could probably be put to better advantage with a larger breeding flock or for ensuring that ewes are in fit condition at mating and lambing. A more logical solution of the marketing problems created by very seasonal production would be refrigerated storage rather than storage on the hoof which is essentially what is achieved by the retention of prime lambs for later slaughter. Moreover this double fattening does not improve the quality of the resulting carcasses.

Butchers do not favour refrigerated storage, arguing that it is expensive and that once lamb goes into store it deteriorates in quality and will no longer receive the premium it attracts when it is sold fresh. This last point is debatable for, unlike beef, lamb does not deteriorate with deep freezing and a milk-fat lamb, provided the carcass is properly conditioned before freezing, could be much better eating than an 8–9 months lamb that has come off a root break. This was the conclusion of a limited study undertaken at Newcastle University in 1970.

One cannot but suspect that the antipathy of the meat trade towards refrigerated storage of lamb owes something to self-interest. With lamb having its strongest consumer appeal in the summer it is very much in a butcher's interests for him to have a buyers' market when there is a firm retail demand. The seasonal drop in prices realised for lamb in the fatstock markets is very seldom accompanied by commensurate price reductions in butchers' shops.

It would not be a question of refrigerated storage for a massive proportion of July–September slaughtering. If no more than one-fifth of present levels of summer offerings was put into store for release in an orderly way later in the year when lamb is in short supply this would increase the prices buyers would have to pay for lambs over the summer period. Because farmers have enjoyed price guarantees that have cushioned them against low prices for fatstock for over forty years there has been little producer pressure for more orderly marketing. However the prospects are that storage of a proportion of summer slaughterings for lamb will become a reality but ironically for the wrong reason. Because of French insistence on very high prices for sheep-meat in EEC there is a real danger that lamb will be so expensive in Britain, relative to alternative meats, especially table poultry, that there will be consumer resistance which will lead to intervention buying and storage of some of the annual crop of lambs. In this event we might witness the nonsense of cheap lamb joining cheap butter on the wrong side of the Iron Curtain. Fortunately these fears have not yet been realised.

Unquestionably the retention of lambs for sale at higher weights has been encouraged by the grading standards and price schedules that have been operative in Britain under the various certification schemes we have had for fatstock since the introduction of guaranteed prices. In the war years, when meat was strictly rationed, the Ministry of Food naturally put the emphasis on weight without much regard for quality and elements of this have persisted in meat grading, and in the case of lamb this has been to the detriment of demand. It was only in 1981, nearly thirty years after the end of meat rationing, and largely due to valuable work by MLC on carcass classification in relation to market requirements, that worthwhile price differentials were established for leaner lambs up to a dressed carcass weight of 20 kg. With this change there is no longer much to be gained from holding gradable lambs in the 35–37 kg liveweight range for further fattening. This is about the target weight at weaning for twin lambs by a Suffolk ram out of a cross-bred ewe.

This does not mean that finishing of store lambs will be on a greatly reduced scale for there will still be a lot of lambs, mainly from hill and upland farms, that will require additional feeding after weaning before they are ready for slaughter. Here one has in mind not only pure-bred wethers of the hill breeds but also the Longwool crosses with these hill breeds. A Halfbred of Mule wether is an excellent sheep for the break feeding of autumn and winter forage crops and its meat has a better flavour than that of a Suffolk-cross lamb which seems to acquire what can perhaps best be described as a rather tallowy flavour from about five to six months of age.

There will continue to be many, often predominantly arable farmers with no breeding ewes, who will either have some crop residue like sugar beet tops or special forage crops that have been grown as a break between cereals who will continue to be buyers of store lambs in the late summer. However they may find it increasingly difficult to pick up cheap lots because many marginal farmers are now electing to finish their store lambs on forage crops such as rape or the so-called Continental stubble turnips which have been grown as part of a grassland improvement programme. Among other things this practice gives farmers some protection from the vagaries of the store lamb market.

SHORT-KEEP STORES

The finishing of store lambs can be classified into two main classes, namely short-keep and long-keep. In a sense it is an artificial classification because there is no distinct line of division but nevertheless it is useful in that one is able to relate a given type of store to an appropriate feeding regime which in its turn can affect cropping programmes. Short-keep lambs are normally marketed within 8–10 weeks of weaning and very often they will be finished on the farms where they have been bred. This is likely where a farmer has been pushing stocking intensity up to its practical limit for under these conditions it is often necessary to adopt early weaning so as to reduce competition between the ewes and lambs for the best grazing once lambs are only marginally dependent on their mothers' milk. Usually pasture is the principal food and clovery aftermaths, following a hay crop, are justifiably held in high esteem. Apart from the likelihood of freedom from round worm infection, clover has special virtues as a fattening food. The only drawback to using such grazing is the danger of prejudicing a clean field programme for the following year. This risk will be lessened by dosing the lambs before they go on to the aftermath which is a routine treatment that should always be adopted when weaned lambs go on to clean grazing. Another precaution with bought-in lambs of unknown history is an injection with an antiserum for pulpy kidney when they go on to lush grazing. Home-grown lambs that have been vaccinated previously should be safe from this hazard.

There are several suitable crops for feeding short-term stores including combinations with rape and the increasingly popular stubble turnip which with summer feeding will not be sown in stubbles but more commonly after a half-fallow. If rape is the preferred forage Giant Essex is recommended because NIAB trials show that it is a high yielding and very leafy variety. Rape sown in a mixture is a safer crop than rape sown alone because it will be slow in ripening in a growthy season and until it does it is not really a safe crop for lambs. It is deemed to be ripe when the leaves have a purplish tinge which should be the result of maturity and not nitrogen starvation that has reduced the size of the crop which is not worth growing unless it is a full one yielding at least 20 tonne of leaf and edible stem per hectare. Suitable companions for rape sown at $1 \cdot 0$–$1 \cdot 5$ kg/ha are Italian ryegrass (12–15 kg/ha), soft turnips (0·25 kg/ha) or mustard (0·5 kg/ha). Their inclusion adds a little variety to the grazing and also makes rape a safer crop for lambs especially in very sunny weather when rape can produce a photosensitive condition if it is not sufficiently mature.

It is advisable to sow a crop for August–September use in two breaks with the first going in about the middle of May and the second about mid-June. It is possible to seed up to mid-July for October–November use but later sowings are not recommended except in very mild districts like Pembrokeshire or Cornwall. Though one can sometimes get a reasonable crop from late July sowing it usually suffers from frosts and the depredations of pigeons. There is also the likelihood on heavy land of very muddy conditions which not only aggravate wastage of the crop but are detrimental to lamb thrift. There is much to be said on this account for direct drilling of rape, kale or turnips into a turf that has been burned off with paraquat because with firmer conditions underfoot there is much less poaching.

Stubble turnips can be sown from the middle of May through to August (when they can usually be truly described as stubble turnips) to provide lamb feed from late July through to December. They are a better proposition than rape in the late autumn because they are less vulnerable to early frosts and the attentions of pigeons. Many farmers claim that they are a better proposition than rape at any time since yields on a fresh crop basis, and also of dry matter, are at least 40 per cent higher than they are for rape.

Both crops should be break-fed to reduce wastage. One of the advantages of turnips is that rather bigger breaks can be given with less frequent moving of the netting. It is recommended that the lambs should, if possible, have a run-back on pasture. If this is not feasible hay should be provided, preferably in covered racks, because rape with 14 per cent dry matter and turnips with 9 per cent dry matter are both very succulent feeds. Some trough food is advisable but it need not be very heavy: 0·25–0·30 kg/head/day of whole barley or oats or a mixture of the two is an adequate allowance with rape but some added protein is advisable with turnips. This is also an appropriate level of supplementation for lambs at pasture especially in late September when pasture normally starts to deteriorate in feeding value.

It is reasonable to expect that a good crop of rape with a grass run-off and a cereal supplement will provide about 2,500 lamb days per hectare. Assuming that at the starting point of feeding the lambs are weighing 33 kg and the aim is to take them up to 38–40 kg liveweight over a period of 6–7 weeks then the provision would be about 50 lambs per hectare. It is reasonable to expect them to gain nearly a kilogram per week in the late summer but later in the year it is unlikely that this rate of increase will be achieved. Stubble turnips provide appreciably more grazing than rape, either alone or in one of its combinations. A reasonable budget figure is seven weeks' grazing for 80 lambs from 1 hectare of turnips, or in other words 1 hectare will provide for approximately 4,000 lamb days.

LONG-KEEP STORES

Long-keep stores are those finished in the winter to meet the trade for hogget mutton. As was pointed out previously it is not easy to establish where lambs end and hoggets begin so it is safer to think of lambs that qualify for the long-keep category as being any that are marketed from 4–8 months after weaning. The basis of their feeding is usually a specially grown crop such as yellow-fleshed turnips, cabbage, kale or swedes or a crop residue such as sugar beet tops or the remains of a harvested cabbage crop. There is little point in giving an estimate of the contribution that cabbage residues can make. In a good year from the viewpoint of cabbage sales there could be no more than a few hundred lamb days per hectare. In a glut year folding by lambs, as an alternative to discing in, could provide up to 4,000 lamb days per hectare which is the output one could expect from a crop grown specially for feeding sheep.

Sugar beet tops, which are normally available from October to December, are reckoned to provide 1,500–1,600 lamb days per hectare or sufficient keep along with hay and concentrates for 35 lambs per hectare for 6–8 weeks. This is equivalent to a half-crop of marrow-stem kale or yellow-fleshed turnips which are both useful crops for the late-autumn to early-winter feeding period. A budget figure here is 60–70 per hectare for a seven-week feeding period which should result in liveweight gains of the order of 5–6 kg with a box feed allowance of about 0·5 kg per day and continuous access to hay.

Swedes are the most popular of all the root crops grown specially for sheep feeding. They are usually folded off over the inclusive period December–March and they have their greatest importance north of the Humber. A reasonable crop will yield 60–80 tonne per hectare and with folding, which is the only method of feeding that can be contemplated, there will be a utilisation factor of about 80 per cent. A hectare of swedes should, with the necessary supplementary foods, provide keep for about 80 lambs over a period of nine weeks, i.e. about 5,000 lamb days per hectare. The concentrate supplement fed at a rate of about 0·5 kg per day should have approximately 20 per cent of soya meal or its equivalent because swedes are essentially a carbohydrate food. The expectation is an average gain of about 0·8 kg per week for cross-bred lambs from a weekly intake of 50 kg swedes, 3 kg hay and 2·5 kg concentrates. The gross output from a hectare of swedes would be about 1,000 kg of liveweight with a good crop and feeding sheep that have the scope to put on weight.

The feeding of beet tops or the leavings of a cabbage crop is something for nothing in that the manurial values from folding are approximately equal to those of the ploughed-in residues. The principal costs are labour, supplementary food and temporary fencing. Generally a farmer will make money if he is able to sell the fat hoggets for more per kilogram of liveweight than he paid for them. The situation is rather different for a feeding crop such as swedes which is grown especially for the sheep except that one should give some value to a change crop in the rotation. Even so the swede crop should first justify itself as a cash crop which is marketed through fattened hoggets. It goes without saying that it has to be grown as cheaply as possible and this means that the days of hand hoeing and singling are numbered. Beyond the effect on a farmer's pride, it does not matter if yellow-fleshed turnips or swedes are no longer the cleaning crop they used to be because most weeds can be controlled by one or other of the wide range of herbicides that are now available. The emphasis must be on cheap production with precision drilling, pre-emergence spraying and steerage hoeing to minimise labour costs. On the other side of the equation it is vitally important that the price paid for store lambs and the sale price of the fattened hoggets are such that there is sufficient of a feeder's margin to cover costs and leave a modest profit. Herein lies the lottery of feeding store lambs against an uncertain market background.

INDOOR FINISHING OF LAMBS

In those far-off days of the sixties when it was possible to buy barley for just over £20 per ton and barley beef was a profitable undertaking, there was some interest in the application of similar methods to the finishing of store lambs, particularly in late autumn when pasture becomes mushy and rates of gain deteriorate. Among possible advantages there was the bonus from taking tail-end lambs off pastures of a reduced deposition of worm eggs to create problems for the following spring. A number of trials were undertaken at the University of Newcastle and elsewhere to determine whether the use of cheap barley in a feed-lot system for finishing lambs was a viable proposition and an alternative to the growing of special crops for this purpose.

The general conclusions derived from the Newcastle work were as follows: It is important to treat lambs for worms when they are introduced into the feed-lot, especially if they are tail-end lambs that have not been thriving on pasture. It is important, too, that lambs are conditioned to concentrate feeding before they are penned, otherwise there is a danger of digestive upsets with greedy lambs or of a check in those lambs which do not readily take to trough feeding. The feeding of new season barley tends to cause digestive upsets but these are minimal when a mixture of oats and barley is fed together with a protein supplement. Cereals should either be fed whole or rolled but not in a meal form.

It is advisable to offer good-quality hay *ad lib.* throughout the feeding period but to ration concentrates on a gradually increasing scale until about three weeks after penning when the hay and concentrates can both be offered to appetite. Over a 6–8 week feeding period hay and concentrates will make a similar contribution to dry-matter intakes. Down or Longwool-cross lambs appear to be the most satisfactory under feed-lot conditions in that they settle quickly in this new environment, unlike hill-bred lambs which tend to be nervous under restriction. If the lambs are enclosed in buildings it is essential that these should be well ventilated. Otherwise there are considerable risks of pneumonia in the characteristically humid conditions of the late autumn.

Cross-bred lambs of approximately 30 kg liveweight in a medium store condition are capable of making gains of approaching 0·2 kg per day and they should be fit for slaughter after approximately seven weeks at 38–40 kg liveweight. Hill-bred lambs will make appreciably smaller gains and 0·1 kg per day would be satisfactory with this class of lamb.

RETURNS FROM FEEDING STORES

These are highly variable not only from year to year but, as one would expect, from farm to farm. Over the period 1971–9 the average gross margin per lamb on MLC recorded farms was £2.21 with a range from £1.12 in 1971 to £3.26 in 1977. However when margins are recalculated to take into account the effects of inflation the average gross margin is £3.74 with a range from £6.21 in 1973 down to £1.72 in 1979. The general picture since the early seventies is that gross margins have not increased commensurately with the substantially increased cost of store lambs. In the MLC sample the average cost of a store lamb in 1973 was £8.46 and the gross margin was £2.61 but in 1980 the margin had fallen to £2.00 against an average purchase price of £21.70; 1980 was not atypical because 1979 was an even more disappointing year for lamb fatteners. However, 1981 results with store lambs costing £24.16 apiece and a gross margin averaging £4.17 was a much happier one largely because of the interim agreement on a sheep-meat

Table 11. Results from lamb-finishing enterprises 1980–1

	Average (£/lamb)	Top third (£/lamb)
Lamb sales (164·8p/kg d.w.)	30·33	33·44
Less purchase price		
(75·5p/kg l.w.)	24·16	23·55
Feeders margin	6·17	9·89
Variable costs		
Feed & forage	1·70	1·57
Other costs	0·30	0·29
Total variable costs	2·00	1·86
Gross margin per lamb	4·17	8·03
Gross margin per hectare	232·7	487·42
Physical performance		
Average l.w. at start (kg)	32·0	31·5
Average l.w. at slaughter (kg)	40·0	41·5
Gain per lamb (kg)	8·0	10·0
Gain per week (kg)	0·7	0·8
Average dead weight (kg)	18·4	19·5
Feed cost per kg of gain (p)	21·3	15·7
Mortality	2·1	1·4
Stocking rate (lambs per ha)	55·7	60·7

regime for EEC that was finalised in 1980 after long negotiations. Hopefully this will bring greater stability to the financial returns from finishing store lambs.

The figures quoted above are averages covering the good, the bad and the indifferent. A rather different picture is provided by the top third in the sample of 111 lamb-finishing enterprises that MLC recorded for 1980–1. Details are given in Table 11. The striking feature of this analysis is the doubling of gross margins, both per lamb and per hectare, by farmers in the top category as compared with average performers. No one single factor is mainly responsible for the superiority of the top third. As is so often the case in situations like this it appears to be the result of contributions from several sources, for instance cheaper procurement, better marketing, lower food costs, higher rates of gain, higher stocking intensities and lower mortalities; in other words, better management. One can see the importance of this when it comes to assessing profit. It is unlikely that there would be very great differences in average fixed costs for the two categories but even if one puts a figure of £50 per hectare on to the fixed costs of the top third there will be at least a £200 per hectare advantage in profit to make store finishing a not unattractive proposition to the farmer who does the job as well as it should be done.

Chapter 8

BREEDING, SELECTION AND RECORDING

SELECTION OBJECTIVES

It is impossible to deal comprehensively with the problems of flock improvement in the space of a chapter, especially under British conditions, with wide environmental demands on the various classes of sheep. Scottish Blackface breeders have vastly different aims from those of Suffolk or Border Leicester breeders, whose sheep are important for crossing purposes. Then there are the self-contained lowland breeds, like the Romney, which again have different functions.

In some respects, this division of functions between various types of sheep simplifies breed improvement, for if we attempted to incorporate all desirable qualities of sheep for the full range of environments into one breed the task would become insuperable. It is difficult enough for species, such as the pig, which are kept in controlled environments. In the Large White, for instance, breeders are selecting for mothering qualities, growth rate, food utilisation efficiency and carcase qualities; but think of the complexity of their task if they had to consider wool and adaptation as additional selection aims. Indeed, there is a growing school in the pig industry that believes the example of the sheep industry should be followed, by adopting purposeful cross-breeding with breeds which have been selected for a specific function.

The starting point in any breeding programme is a clear-cut definition of objectives and in this one must be realistic—for instance, the Roman nose so highly regarded in some breeds may be no more than predisposing to jaw deformities. It is important to remember that every additional character in a selection programme materially reduces the progress that can be made in respect of other characters under consideration. Unfortunately, there is no very clear-cut definition of selection objectives in most sheep breeds. Individual excellence, which may be no more than the breeders' concept of a particular breed type, is the main basis of selection and, generally speaking, there is very little done in the way of objective measurement of traits of unquestioned economic importance. The danger of relying on type appraisal alone is very well reflected by the results of a trial undertaken a number of years ago with Southdown rams in New Zealand. Two groups were selected for cross-breeding to produce fat lambs. The first group belonged to the 'stud' classification which covers rams considered suitable for pure breeding. The second group consisted of discards not considered suitable even for fat-lamb production. The resulting progeny test disclosed no significant difference between the two groups of rams as fat-lamb sires and, in fact, the best ram belonged to the reject group.

Selection objectives must be defined in strict accordance with the functions of a breed in the environment in which it has to live. In hill breeds, adaptation to a difficult environment—which is a complex of not very well-defined characters—is all-important. It includes the capacity of the lambs to survive hazards confronting them immediately they are born, foraging ability, an instinct to protect themselves in snow storms, and a capacity in pregnant ewes to give birth to viable lambs despite the loss of a high proportion of their body weight in the latter stages of pregnancy.

Because it is impossible to separate the various components of this adaptation complex, one can only select for the power to survive and reproduce in a particular environment. There seems to be no point, indeed a positive danger, in buying rams of a hill breed from a flock maintained under what are essentially lowland conditions. Yet this sort of thing happens, and with it there is a danger of breeding sheep too large for their environment. Size seems to be an important adaptative quality, because small sheep have a comparative foraging advantage over bigger sheep under difficult grazing conditions.

This does not deny the importance of selecting for heavier-fleeces or heavier weaning weights, because these are economically important characters, but it is prudent to select in the appropriate environment so that adaptive qualities are safeguarded. Work at Bangor with Welsh Mountain sheep, kept under their natural conditions, has shown that it is possible to effect improvements in growth rate of lambs even though conditions are not optimal for expressing this character.

At the other end of the stratification range, where Down breeds are used to produce a slaughter generation, little emphasis need be placed on mothering attributes or fleece weights. The main requirements are rapid rates of gain and good carcasses. Unfortunately, though it is easy to say that a sheep must transmit good carcass attributes, these are ill-defined but it is important to have a high proportion of total carcass weight in hindquarters, good development of the eye muscle, and a relatively thin but complete fat cover without large tallowy patches. Such carcass features are greatly influenced by the degree of fattening which in its turn is influenced by the inherent nature of the animal, the level of nutrition to which it has been subjected, and the duration of the feeding period. Here we see the interplay of nurture and nature in determining the end result, and this constantly complicates the task of the breeder.

STRENGTH OF INHERITANCE

This leads to a consideration of heritability or strength of inheritance. When the performance of a parent is a good indicator of its breeding value, the character in question is said to be strongly inherited. Conversely where there is a poor correlation between the performance of parent and offspring the character has a low heritability and it is said to be weakly inherited. Examples of traits with relatively high heritability estimates are fleece weight $(0 \cdot 30-0 \cdot 45)$ and fleece quality $(0 \cdot 40-0 \cdot 70)$. Generally reproduction traits are weakly inherited, for instance estimates of the heritability of both litter size and milk yield range from $0 \cdot 10$ to $0 \cdot 20$ while carcass composition and conformation occupy an intermediate position with values ranging from 0·25 to 0·35.

A knowledge of heritability values is of basic importance in formulating effective selection programmes. If we are concerned with a character of very low heritability, this does not mean that there will be little flock variation in respect of this character. To the contrary, for there will be variability attributable to genetic interactions, and, especially, differential effects of environmental factors, such as incidence of infections, type of birth (twin or single), planes of nutrition and so on. With very weakly inherited characters one can select, generation after generation, what appear to be the outstanding members of a flock and end up with animals that are no better than the original foundation.

Where a character is strongly inherited and it can be measured accurately in the living animal, it is possible to make substantial and rapid genetic progress, for one can exploit the technique known as performance testing which has such importance in pig and beef cattle improvement schemes. In pigs both weight for age and economy of food utilisation are relatively strongly inherited, and so is the depth of fat cover over the eye muscle which can be fairly accurately measured by ultrasonic methods. These characters are important economically. Performance testing permits a large number of young animals to be screened at a comparatively low cost so that only good prospects go on to the more expensive progeny test, which gives the most reliable estimate of an animal's breeding worth, especially in respect of those characters which cannot be assessed by performance testing.

To transfer these thoughts to sheep, let us assume that by early weaning at say, five weeks, and supplementation with concentrates from the earliest possible age in order to reduce maternal effects (largely attributable to the differential milking capacity of dams) we are able to establish that the sixteen-week weight of Suffolk rams is highly correlated with the sixteen-week weights of their cross-bred fat lamb progeny. We would then be in a position to make reasonably rapid progress in breeding Suffolks with superior weight-for-age performance.

VALUE OF RECORDING

The sheep industry is at a considerable disadvantage as compared with both dairying and pig production in respect of records of performance that can be analysed to give reliable estimates of strength of inheritance of economically important traits and also of genetic correlations between traits. Analyses of a sequence of annual records can establish whether early records are good indicators of life-time performance. For instance, has fleece weight and the quality rating of wool at a sheep's first shearing a close relationship with its subsequent performance? The answer is yes if there are no environmental upsets and this not only allows a breeder of a dual-purpose sheep like the Romney to make decisions as to retention or culling of sheep, both rams and ewes, at the yearling stage but also obviates the need to take subsequent records for these traits. This is important because much effort has to be put into recording and it is counterproductive to keep records that not only have a very limited value but also compete with the collection and interpretation of more important records.

NEW ZEALAND'S 'SHEEPLAN'

Simplicity is the keynote of 'Sheeplan', the pedigree flock improvement programme of the New Zealand Ministry of Agriculture that owes a great deal to the initiative of progressive pedigree breeders who had laid the foundation of the programme in their individual efforts to improve their sheep. It is a large-scale operation and in 1978 there were just on 900 breeders involved with a quarter of a million breeding ewes. These are the records that are taken, the first of which is compulsory:

(a) number of lambs born and/or reared;
(b) weaning weights which are indicative of milk yields and growth rate of lambs;
(c) live weight of hoggets, which may be taken in the autumn, winter or spring;

(d) hogget fleece weight and quality (it is not usual for fleece recording to be undertaken with mutton breeds because the emphasis in selection with these is on growth and carcass traits).

The New Zealand scheme is only valid for within-flock comparisons. Inter-flock comparisons are ruled out because of differences in management that may mask any genetic differences. Most of the flocks in Sheeplan are relatively large. The average flock has 280 ewes, and about one-quarter of all flocks have more than 400 ewes while there is also an appreciable number with more than 1,000 ewes. This means that meaningful within-flock comparisons can be made; for instance a 500-ewe flock could have most of its ewes mated to promising ram lambs, out of selected ewes, in a progeny test to discover recruits for the team of elite rams of proven ability that are mated to the top ewes. These will not only produce replacement ewe hoggets but also rams either for testing in the home flock or for sale.

Breeders are responsible for collecting the raw data for the individually identified sheep in the flock and these are processed centrally by computer and corrections are made for type of birth (twin or single), age of dam, etc. The processed data are returned promptly to the breeder so that he can make the necessary decisions on retention or culling. The processed data include estimates of breeding values which predict the average performance of progeny of rams and ewes based on their own performance. For instance if a ram has been given a breeding value of 0·2 kg for fleece weight it means that he is not only appreciably above the average rating for the flock (which is zero) but also that if he is mated to average ewes their progeny would be 0·1 kg above average.

RECORDING IN BRITAIN

MLC has introduced a recording scheme for pedigree flocks in Britain covering individual ewes over successive years. The emphasis in this is on fertility, prolificacy and mothering ability as measured by the number and weight of a ewe's lambs at eight weeks of age. The first report of the scheme listed 259 flocks, covering 32 breeds with 26,262 ewes put to the ram which gives an average of only 100 ewes per flock. These figures reveal something of the structural weakness of pure-bred sheep breeding in Britain. The report records only eight flocks with more than 400 ewes which is about the minimum for a closed flock in a breeding plan while 63 per cent of the flocks had less than four rams, including 15 per cent with only one ram. All but one of the eighteen recorded Border Leicester flocks have less than 100 ewes and here the one exception boasted only 150 ewes and three rams. There were seventy-five Suffolk flocks, the most numerous group, of which only eighteen had more than 100 ewes. Clun Forests had the best breeding structure of all the breeds in that only one of the eleven recorded had less than 100 ewes and there was an average of 200 ewes and five rams per flock.

It is not inferred that flock records are unimportant where numbers are small, to the contrary, but that they cannot be used to such good effect either by the breeder or the prospective purchaser of breeding stock as they are in the large recorded flocks such as those in New Zealand. Even after preliminary culling mainly at or prior to weaning, a typical Romney or Perendale breeder in New Zealand could have 100 two-tooth rams with breeding values for wool, weaning weight, yearling weight and prolificacy from which to choose twenty tups for his own use. If he is not satisfied with home-bred sheep or feels there is a need for an outcross he can go to fellow-breeders who are also recording to make up the numbers. Possibly he will have 300 ewe hoggets also with predictions of their breeding value and a replacement quota of two-thirds this number. He is able to select with greater confidence than he could in pre-Sheeplan days when appearance was virtually the only criterion. The same applies to commercial farmers when they are buying their annual quota of flock rams. In the old days it was the size of the rams and their shapes, usually modified by judicious trimming subsequent to early shearing, that appealed to buyers. Now they have something much more meaningful on which to base their purchases and this also benefits the sheep industry at large.

DEVELOPMENT OF A BREEDING PLAN

Improvement in breeds like the Clun or the Romney, where there is a high proportion of pure- as opposed to cross-breeding and where there are well-defined selection goals such as mothering qualities, growth rates and fleece weights, is relatively straight-forward as compared with crossing breeds such as the Border Leicester or the Suffolk. Here merit lies less in pure-bred performance than in the performance of the cross-bred. By way of illustration, the development of hybrids has brought a totally new dimension into maize growing but yields of the parent lines of hybrids are usually very poor. Their merit rests primarily in the way they combine to produce superior hybrids and they have been selected for this function. One cannot help but wonder when looking at Bluefaced Leicester rams, which are anything but handsome sheep, whether the undoubted quality of their cross-bred is not due to something akin to the hybrid maize situation.

We do not have this problem in breeds like the Clun or Scottish Blackface where the primary emphasis in selection rests on improvement of performance of pure-breds and the performance of cross-breds is a secondary consideration. Here, provided breeders have the will to record and use records and the necessary numbers, it is possible to develop breeding and selection programmes that will have reasonable prospects of success. Before examining any plan of action they should consider some of the factors that control rate of progress towards selection goals. The principal determinants are as follows:

(a) the number of selection objectives and their compatibility;
(b) the selection differential;
(c) the accuracy of measurement;
(d) the generation interval.

These require further definition and some explanation. The greater the number of characters that are included in a selection programme the slower will be the rate of progress in any one direction. For instance much less progress will be made towards greater fleece weights if there is an additional requirement of increased fibre fineness. Apart from having two objectives, fine-woolled sheep tend to have lighter fleeces than strong-woolled sheep so there is a measure of incompatibility between objectives. Sometimes the reverse holds, in other words there can be a positive as opposed to a negative correlation between traits; for instance lean, fast-growing pigs are more economical converters of food than slow-growing fat pigs. In consequence there is no conflict between objectives if a breeder seeks to speed up growth rates, reduce back fat and improve economy of food use all at the same time. It is unwise to try to fulfil too many aims in a breeding plan while priority should be given to traits that have the greatest economic importance.

The selection differential for a given character is the difference between the average value for the breeding population and that of animals selected to be parents of the next generation. It is important to remember that this differential, possibly expressed in terms of units of liveweight gain or weight of fleece, depends on more than genetic factors. High producing animals tend to have had less in the way of environmental disadvantages than their less productive contemporaries. This is where the knowledge of heritability values is important for this will provide an estimate of the proportion of recorded differences that is attributable to additive genetic factors and is therefore transmissible. For instance, we might have two groups of ewes within the same flock and under the same general management whose average wool clips as shearlings differed by 500 g per ewe. However, we could not expect their progeny, sired by the same ram, to differ by 250 g at first shearing which would be the case if the difference were completely heritable. Fleece weight has a heritability factor of about 0·4 and this means that the probable differences in shearling fleece weights between the two progeny groups would be about 100 g.

Some measurements can be taken with considerable precision—for instance weight of fleece—but this relates to only one year in the working life-time of a sheep. However, scientists responsible for Sheepplan in New Zealand have established that the shearling clip is a reliable indicator of life-time performance and it is reasonable to suppose that this will be the case in British flocks provided, of course, that a sheep does not suffer some serious upset that impairs future production. A reliable indication of an animal's future performance that has been obtained at an early stage of its productive life is of great importance in a breeding programme because it means that decisions on retention or culling need not be unduly delayed. This in turn will lead to a shortening of the generation interval which has a profound influence on the rate of genetic progress. For instance if an improvement of 480 g per generation is made in weaning weights and the generation interval is six years, that is not as good a result as a 360 g gain made in a flock with a four-year generation interval.

The breeder, in a sense, is on the horns of a dilemma particularly if he is dealing with traits that are greatly influenced by environmental effects, such as numbers of lambs born per ewe. If he waits until five lambings have been completed before he decides to save flock replacements he has a reliable assessment of life-time prolificacy but it is obtained expensively in terms of time and recording effort. If he makes his decisions after three lambings per ewe he will lose something on accuracy of assessment but this could be more than compensated for by the shortening of the generation interval.

Dr Maurice Bichard encountered this problem when he controlled a breeding project with Clun Forest sheep at Cockle Park over the period 1960–74. The principal aim was improvement in mothering attributes, i.e. prolificacy, viability of lambs and milk yield of ewes. In the earliest years of the trial when there was an emphasis on life-time performance in order to establish basic parameters the average age of dams of flock replacements was six years. Later with the analysis of accumulated records he established that three lambings (including that as a yearling) provided a reasonably reliable estimate of life-time performance. As a consequence he was able to reduce the average age of dams contributing flock replacements from six years to four years. Possibly there was some loss of accuracy in selection but this was more than compensated for by one-third reduction in generation interval.

BREEDING PLAN FOR A EWE BREED

Here we are considering breeds like the Clun and Romney, and though they are dual-purpose breeds the emphasis in Britain must be firmly placed on meat rather than wool. Although wool should not be neglected it would come into consideration in the selection programme only on an other-things-being-equal basis because wool accounts for only 10–15 per cent of flock income with most of the balance coming from lambs. The records of greatest importance are (1) pedigrees of individual animals, (2) number of lambs born and reared at each lambing, (3) date of birth of lambs and, (4) weaning weight of lambs.

If it is intended to maintain a closed flock, which may be a decision forced on a breeder because there are no other breeders who are exerting similar selection pressures towards the same objectives, a safe provision is not less than 400 ewes and fifteen rams. This will give the necessary scope to maintain a low coefficient of relationship within the flock without any serious limitations on selection arising from efforts to avoid close matings. If there are fellow-breeders with the same objectives and policies the owner of a small flock can safeguard his position by appropriate purchases of rams that will provide an outcross without detriment to the general performance of his flock. Before a flock is closed the breeder is advised to widen his genetic base using unrelated rams from several sources, preferably those with no blood relationship with his flock.

It is suggested that all replacement ewes should be mated as lambs. Among other things this will appreciably reduce the age at which a reliable assessment of a ewe's breeding capabilities can be made. It should be possible to decide at the three-year stage whether ewes should be discarded or retained to breed flock replacements. The expectation, even with reasonably good nutrition and freedom from worm infections is that only about two-thirds of the ewe lambs will hold to the ram.

The main breeding flock, i.e. two-tooth ewes and older, should be mated to ram lambs that obviously must be well grown if they are to cope with 25–30 mature ewes apiece when they are only seven months old. They would be selected from the best ewes in the flock judged on performance and these would normally be four- and five-crop ewes. In order to guard against inbreeding it is recommended that not less than twelve unrelated tups should sire flock replacements (both male and female) in any one year and that these sires should be turned over every year.

It does not necessarily mean that such a prodigal use of tups will be expensive, certainly not in the case of somebody who is breeding rams for sale, because after limited use as lambs their value as shearlings will not be diminished. The breeder will be advised to retain a small contingent of older rams to mate with the ewe lambs because experienced rams appear to be more effective than tyros in effecting conception when they are joined with ewe lambs.

Possibly the most demanding task facing the breeder will be the interpretation of the records because allowances have to be made in an assessment of weaning weights because of differential growth rates due to sex, age of dam, size of litter and numbers reared per ewe. Unless he is in the unlikely position of having his own computer and skill in handling data he would be strongly advised to contact MLC or a government livestock advisory officer to see if he can have the benefit of a data processing assistance similar to that provided by Sheepplan in New Zealand. Otherwise the best he can do is to use his judgement when interpreting individual records.

IMPROVEMENT OF CROSSING BREEDS

A flock improvement plan for Longwool breeds like the Border Leicester has to take into account that merit depends not on pure-bred performance alone. It also depends on the contribution the breed makes to the production of better cross-breds, particularly in respect of mothering qualities, and assessing this would constitute a forbidding task beyond the resources of any one farmer. The situation is somewhat less demanding in breeds that are used primarily to produce a terminal generation, for here the principal objectives are improved growth rates and depth of fleshing which are measurable at 16–20 weeks of age. It is reasonable to expect that if a ram excels in these respects his cross-bred progeny will inherit some of his superiority, but one cannot be certain of this except by progeny testing which will involve carcass assessment.

In the sixties we established a small flock of Suffolks on the University's Nafferton farm, which had a 500-strong Scottish Halfbred flock. A programme was devised where young Suffolk rams, some home-bred and some purchased, were mated to Halfbred ewes with the intention of rating them on the basis of growth rate and conformation of progeny. The best ram was used in the next year on the pure Suffolk flock partly to produce more rams for testing. We quickly ran out of elbow room, for animal improvement is essentially a numbers game and within a short space of time we found that the best of our tested rams had a high proportion of close relatives in the flock. Our fifty Suffolk ewes, a very typical flock size in this breed, needed a multiplication factor of at least ten to give the necessary scope for the breeding and selection programme that was envisaged.

Regretfully the project was abandoned. It is encouraging to learn that the East of Scotland College of Agriculture has recently embarked on a similar plan of action but that wisely they are making it a co-operative venture with participation by several breeders. One hopes that it will succeed, not just in respect of better Suffolks but in demonstrating the benefits of co-operation in breed improvement schemes where individual flocks are too small to support independent programmes.

It is to be hoped too that the East of Scotland programme puts a particular emphasis on the cut-out value of carcasses and on area of eye muscle which is such an important component of the lamb chop. The Suffolk may have a better growth rate than the Texel but reputedly the Dutch breed is a clear winner in depth of fleshing where it really matters, and this represents a challenge that Suffolk breeders must meet if their breed is to maintain its dominance. This point is of particular significance in any expansion of lamb exports to the Continent.

SYNTHETIC AND IMPORTED BREEDS

There have been many attempts in several parts of the world to produce synthetic breeds from a foundation incorporating two or more established breeds. There has been a great deal of interest and considerable activity in this approach to sheep improvement in the past thirty years but it is by no means a novel development. Just on two centuries ago George Culley in North Northumberland was using Leicester rams from Robert Bakewell's Dishley flock on local sheep known as 'mugs', which seems a somewhat inapt name for sheep that have contributed to the unquestioned excellence of the modern Border Leicester.

A hundred years ago, with the advent of refrigerated transport, there were successful attempts to develop dual-purpose breeds to replace the pure Merino in several New World countries. The best known of these are the Corriedale in New Zealand and the Polworth in Australia while the United States produced the Colombia and Romilet. The success New Zealand has had with the Corriedale seems to have gone to breeders' heads because of the twenty-six breeds listed in the Sheeplan report no less than fourteen are new breeds that have been developed in the past forty years. The most important of these parvenus is possibly the Perendale, a Romney–Cheviot derivative which has a very useful role on hill farms that are not really good enough to carry Romneys.

There are two other new breeds that have also achieved prominence, the Coopworth which endeavours to combine the virtues of the Romney with those of the more fecund Border Leicester, and the Drysdale, which is not a synthetic. Rather it has arisen by accident and, like Malvolio in Twelfth Night, it has had greatness thrust upon it. The starting point was a very odd-looking lamb born in a pedigree Romney flock owned by a man with a naturalist's interest in his sheep. Instead of knocking it on the head at birth, as most breeders would have done, he presented it to Dr F. W. Dry, a geneticist on the staff at Massey College who was then working on the problem of unwanted hairiness in Romney fleeces. At maturity it was even more grotesque looking than it was at birth. Not only did it have a grossly hairy coat, not unlike that of a

Scottish Blackface, but it also had horns, which is not a Romney characteristic.

It was mated to Romney ewes with varying degrees of hairiness in an attempt to establish the mode on inheritance of the condition and in the course of a few years Dry had assembled a collection of odd-looking descendants which became something of an embarrassment to the College. He was loth to dispose of any of his sheep which were putting pressure on the available resources and they were also an affront to unsympathetic colleagues who were interested in building up a flock of Romneys that were completely free of the offending hairy fibres. So much pressure was put on Dry that he was forced to arrange accommodation for his surplus sheep with a farmer friend and pay for their keep out of his own pocket. Fortunately at this point they were discovered by a company that had just started New Zealand's first carpet factory and which had been forced to import Blackface wool from Britain to blend with pure Romney wool that lacks the necessary resilience to make a really good carpet yarn. It was established that Dry's oddities had fleeces that were ideal for their purpose and the Drysdale breed was born. The few fortunate farmers who were able to get Drysdale rams at the beginning received a substantial premium for their wool over the prices unblemished Romney wool was realising.

The irony with this rags-to-riches breed is that its originator was a completely dedicated scientist with never a commercial thought in his head. If this story has a moral, apart from a reminder that valuable discoveries can often come out of dirty test tubes, it is that genetic material, be it in the shape of a rare breed or an off-type animal like the progenitor of the Drysdale, should never be lightly discarded because it may, under changed circumstances, have a valuable commercial contribution to make.

It was such a recognition of the need to have a wide genetic basis in developing new breeds that led to the importation of at least two foreign breeds, the East Friesland and the Finnish Landrace which, as pure breeds, have no commercial future in Britain. The East Friesland, because of its milking ability, was one of the four breeds of the foundation that led to the establishment of the Colbred which is the best known of the

British synthetics. Oscar Colburn's pioneering venture was not only courageous but also a soundly based attempt to establish a crossing breed that combined fecundity, milk yield, an extended breeding season and reasonable carcass quality. The fact that it has not yet made a great impact cannot be attributed to its shortcomings but rather to the excellence of the well-established crossing breeds, in particular the Border and the Bluefaced Leicesters.

The Finnish Landrace was used in the development of the Cadzow Improver which was intended for intensive fat-lamb production with environmental controls that included housing, lighting regimes and hormone manipulation to promote out-of-season lambing and production. Possibly it was part of a concept that was launched before its time but the sad fact for its originator is that it did not succeed. It is a very long-term and expensive exercise to develop and launch a new breed that can compete successfully with entrenched breeds and crosses in what is possibly one of the most conservative sectors of British agriculture. The Finnish Landrace is being used by the Animal Breeding Research Organisation in the evolution of a sire line for crossing on hill breeds and it is also being used in Eire in a long-term programme undertaken by the Agricultural Research Institute to increase prolificacy in the local Galway breed.

There have been other attempts to develop new breeds. Henry Fell, a progressive Lincolnshire farmer who has been one of the leading lights in the introduction of the Oldenburg, has produced the Meatlinc and Dr John Owen when he was with the School of Agriculture at Cambridge started a programme of interbreeding Welsh Halfbreds. Neither of these ventures has made an appreciable impact and this is not surprising in view of the competition they have to face from established breeds and crosses. In addition they have the disadvantage of being started on the arable east side of Britain where sheep farming is now a minor activity.

Two mutton breeds that have been imported, the Ile de France and the Texel, have had contrasting fortunes. It does not appear that the Ile de France has much of a future in Britain. Though it has a fairly well developed eye muscle it does not compare with the Texel in this respect and it is a poor second to the Suffolk in growth rate. The Texel, on the other hand, enjoyed an initial boom and early breeders have received very high prices for their stock because of a strong interest in the potential of the breed which many felt would achieve a role parallel to that of the Charolais in beef production. But this interest has now waned and farmers have mainly gone back to their old loves in the shape of the native Down breeds, in particular the Suffolk. However there is another recent sheep importation, also called the Charollais, that is now meeting a sellers' market. It has particularly deep muscles over its hind legs and loin and promises to be valuable in the production of lamb carcasses destined for Continental markets, particularly when mated to hill-bred draft ewes whose lambs normally leave a little to be desired in respect of carcass conformation.

Chapter 9

FACTORS CONTROLLING PROFIT

GENERAL PRINCIPLES

Production costs consist of two parts, fixed or overhead costs and variable or running costs. The difference between the value of output and the variable costs is termed the gross margin which is a useful comparative measure of the performance of an enterprise but not of profit for this also depends on the impact of fixed costs. If these exceed the gross margin then the enterprise is running at a loss.

Fixed costs in farming include interest and bank charges on the investment in land, equipment and livestock, managerial expenses, and the wages of permanent employees, while seeds, fertilisers, purchased feeds, fuel, power, transport and veterinary services come under the heading of variable costs. A farmer has only limited scope for reducing overheads apart from repayment of loan indebtedness. Rent, for instance, may be subject to renegotiation but recent trends have been upward and not downward. He may be able to reduce permanent labour by increased mechanisation but this will add to capital costs and depreciation. His best course of action could either be increased output to give a better spread of overheads or technical changes that lead to higher labour productivity with little need for further investment. A good example of this was the adoption of a quick milking routine and the elimination of hand stripping in dairy farming.

The greatest scope for increasing profitability probably rests on an increase in the value of output without any disproportionate increase in overheads or running costs. This may be effected by increased output, improved quality of output, or a combination of both. There are notable examples where

127

increases in output have been obtained with only trifling increases in costs; for instance the sowing of higher-yielding cereal varieties, improved hygiene to reduce wastage in intensive livestock production and the switch from Dairy Shorthorns to Friesians in milk production.

Sometimes spectacular improvements can stem from one simple action, for instance the application of a fertiliser containing a small amount of cobalt sulphate to land where ruminants suffer from the disease commonly known as pine. Here the difference between normal health and a wasting death can be a matter of 7–8 parts of cobalt per million parts of dry matter. More usually increased productivity, with higher profitability, is not due to a single factor but to the collective effects of a complex of factors that all contribute to a given end. The secret of successful management is the recognition of these essential details and giving them due attention in the organisation and running of an enterprise.

MANAGEMENT PRIORITIES

Until comparatively recently sheep production, unlike dairying or pig and poultry production, suffered from a dearth of reliable data covering the physical and financial aspects of the industry that could be analysed and interpreted to improve efficiency of production. Until the establishment of the Meat and Livestock Commission with its Sheep Improvement Services there was no body comparable with the Milk Marketing Boards or the Pig Industry Development Authority (now incorporated in MLC) that was able to take a synoptic view of the industry and provide guidelines for its future development. This deficiency has been made good and since 1972 there has been a steady flow of useful information such as that given in Table 12.

The top third of the lowland spring lambing flocks had a gross margin advantage over the average of £153.17 per hectare (62 per cent) while the corresponding advantage in upland flocks was £104.99 (41 per cent). An interesting feature of these figures is the impact of the ewe subsidy which not only narrows the difference between the average for all upland farms and that for the top third but also gives upland farmers a higher gross margin per hectare than that recorded for the average lowland farms. The ewe subsidy in 1980 increased the average gross margin on upland farms by approximately £40 per hectare or 16 per cent and on top third farms by 11 per cent.

An even more interesting feature of the table is that in each category the superiority of the top third is the result of better

Table 12. **Physical and financial results per ewe for lowland and upland flocks, 1980**

	380 Lowland flocks		87 Upland flocks	
Financial results (£)	*Average*	*Top third*	*Average*	*Top third*
Lamb sales	38.49	43.65	32.70	38.74
Wool	3.15	3.18	2.38	2.44
Ewe subsidy	—	—	4.05	4.04
Gross return	41.64	46.83	39.13	45.22
Less replacement cost	6.45	4.96	4.60	4.21
Output	35.19	41.87	34.53	41.01
Variable costs				
Concentrates	6.25	5.80	4.06	3.74
Other purchased food	0.45	0.47	0.12	0.09
Forage costs	4.35	4.83	3.02	2.84
Total feed	11.05	11.10	7.20	6.67
Veterinary costs	1.86	1.76	1.21	1.13
Miscellaneous costs	0.95	0.78	0.64	0.63
Total variable costs	13.86	13.64	9.05	8.43
Gross margin per ewe	21.33	28.23	25.48	32.58
Gross margin per hectare	246.34	399.51	256.60	361.59
Physical results				
No of ewes	398	387	441	409
No of lambs born (%)	160	173	139	150
No of lambs reared (%)	141	157	122	136
Lamb mortalities (%)	19	16	17	14
Ewes/ha	11·5	14·2	10·0	11·1
Kg/N/ha	151	181	68	88
Kg/N/ewe	13	12	7	8

performance in practically every aspect of production. For instance on average the best farmers spend slightly less per ewe on concentrates and medicines and they have appreciably lower flock depreciation. They may spend more on fertiliser nitrogen per hectare but because they are efficient in utilising the additional herbage production there is no difference in nitrogen usage per ewe. The factors that make the greatest contribution to superior performance, as the analysis in Table 13 shows, are the number of lambs reared per 100 ewes and the intensity of stocking.

Table 13. Percentage contributions to top third superiority in gross margins per hectare

	Lowland flocks	Upland flocks
Sale price of lambs per head	7	20
No. of lambs reared	33	36
Flock replacement costs	11	4
Feed and forage costs	2	5
Stocking rate	37	25
Other factors	10	10
	100	100
Effect on GM per ewe of 0·1 increase in lambs per ewe	£2.56	£2.33
Effect on GM per ha of 0·1 increase in ewes per ha	£1.89	£2.27
Extra GM of top third per ha	£153.17	£104.99

Obviously if a sheep farmer wishes to make a substantial improvement in gross margin per hectare he would be well advised to pay special attention to improvement of the effective lambing percentage (i.e. the number of lambs reared per 100 ewes put to the ram) and the stocking intensity of the farm. The realisation price per lambs is also important especially on upland farms. Here it is a matter of considerable concern to a farmer that he sells a high proportion of his lamb crop in slaughter rather than store condition or, better still, that he is in the fortunate position of selling fashionable cross-bred ewe lambs, destined to be fat-lamb mothers.

The improvement of lambing percentages does not necessarily involve a disproportionate increase in costs. In fact the figures given in Table 10 suggest a reverse situation with fecund ewes that respond to favourable management such as correct conditioning for mating and for lambing. Another front on which it should be possible to make worthwhile advances as a result of greater care, especially just before and just after lambing, is a reduction in lamb mortality. Even in the top thirds of the two categories mortalities approximate to 10 per cent of the total drop of lambs.

STOCKING INTENSITY AND NITROGEN USAGE

Stocking intensity which is second in importance to prolificacy depends on a number of factors such as size of ewes, timing of lambing, the inherent fertility of the land, and management of pastures and associated forage crops. Among pasture management practices the one that has the greatest impact on productivity is the level of fertiliser nitrogen usage. Even under favourable conditions in Britain the annual output of dry matter from a balanced ryegrass-white clover sward will not exceed 5,000–6,000 kg per hectare. This is about half the figure attained under comparable New Zealand conditions where there is a much longer growing season, especially for white clover with its double function of being a source of nutriment for grazing animals and a source of nitrogen for its companion grasses. The New Zealand farmer is in the fortunate position of having cheap clover nitrogen while his British counterpart has to rely to a considerable extent on bag nitrogen which is not cheap. Nevertheless it is not so expensive that it does not pay to use it on pastures to produce more milk and meat, providing the extra herbage is utilised efficiently.

If there is no severe limiting factor such as a very late spring or a prolonged summer drought there is a straight-line response to fertiliser nitrogen applications up to a level of approximately 300 kg N/ha. The response per unit of N varies according to the quality of land and seasonal conditions but a typical response is 25 kg DM/kg N. In other words an annual application of 200–250 kg N/ha will double the production of dry matter per hectare, and with efficient utilisation double the carrying capacity of grassland. In practice, despite this straight-line response in dry matter production, utilisation expressed in terms of grazing days or liveweight gain per unit area tends to decline after the 150–180 kg N/ha level of application. This is understandable because of the greater competition for available food and the increased fouling of pastures at the higher levels of stocking that are necessary to effect complete utilisation of the additional herbage.

Nevertheless as shown by Figure 4 which relates to farms recorded by MLC in 1979 there is a steady rise in the gross margin per hectare with increasing rates of nitrogen application. There is a bias built into these figures inasmuch as it will generally be the most efficient farmers who will be the heaviest users of nitrogen. Reference to Table 12 will show that the top third of lowland farms has a figure of 181 kg N/ha as compared with 151 kg N/ha for average farms in this category.

Nevertheless it is clear that farmers who give the necessary attention to the management factors that are important in achieving high levels of physical performance can make substantial increases in gross margins per hectare by a judicious use of fertiliser N.

Primarily this improvement will stem from increased stocking and it can be seen from Figure 5, again based on MLC records for 1979, how gross margins per hectare rise steadily in response to progressive increases in stocking intensity

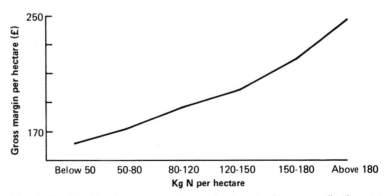

Fig. 4. *Relationship of gross margins per hectare to nitrogen application per hectare of grass on lowland sheep farms.*   MLC

despite an accompanying decline in gross margin per ewe.
Possibly this decline will be less marked now that price
schedules have been changed to discourage the marketing of
heavy over-fat lambs which are favoured by low-intensity
stocking, but nevertheless this decline in gross margin per ewe
remains as the principal challenge in the intensification of
sheep farming on both lowland and upland farms.

Again we see the importance of high management standards, for instance the judicious but not extravagant use of
supplementary concentrates early in lactation to lessen the
pressure on pastures. Flock hygiene, particularly in respect of
the control of internal parasites, has an increased significance
with high-intensity stocking. In fact it can be said quite
categorically that high-intensity stocking of ewes and lambs is
doomed to failure if the simple but essential precautions to
avoid heavy parasitic infections are neglected.

Some of the decline in gross margin per ewe may be due to
the deliberate choice of a small ewe (See Table 10). One can
carry more cross-Welsh ewes per unit area than Scottish
Halfbreds and this could make sense because a small ewe has
an appreciable adaptive advantage with high stocking intensities in that it is better able to satisfy its needs where there is a

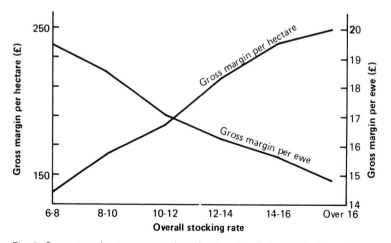

Fig. 5. Gross margins per ewe and per hectare in relation to stocking rate per
hectare.   MLC

pressure on grazing. Nevertheless to give of its best a small ewe must also be very fecund and in this respect the Welsh Halfbred suffers in comparison with its Scottish counterpart, or the Mule and the Greyface.

LABOUR EFFICIENCY

There was a time when a shepherd, once the highest paid of farm workers, was fully employed looking after a flock of 200 ewes. He spent his days, in all kinds of weathers, repairing and shifting the hurdles that were used to limit the breaks of folded crops, turning up ewes to dress their feet if they were lame or brushing out the maggots resulting from blow-fly strikes which could be an all too common occurrence in a humid summer. He could recognise all the ewes in the flock and give their rearing performances just as a good cowman can recite the yields of members of his herd. Remarkably, though he often lacked numeracy, he was generally able to tell if sheep were missing from his flock.

These days have passed and now we expect a shepherd to be a general stockman who looks after a herd of suckler cows as well as 400–700 ewes and to lend a hand at hay and harvest times. It is not that standards of shepherding have declined. The gains have come from streamlining routine operations and from putting the emphasis on labour-saving measures such as pastures instead of folded crops and, if the latter are used, wire-netting instead of wooden hurdles to form the breaks. High on the list of labour-saving measures is the provision of sheep-proof fencing. It is important for many reasons that sheep should remain confined to the field where it is intended that they should be, and a great deal of time can be wasted chasing sheep out of corn crops and patching up holes in hedgerows with hurdles or old iron bedsteads.

The availability of effective remedies and efficient prophylaxis have made a notable contribution to higher labour productivity. The emphasis has now moved from treatment to prevention; for instance the elimination of foot-rot from a flock by the systematic use of formalin baths and field hygiene. Modern dips can also reduce the nuisance of blowflies and headflies, and vaccination programmes have supplanted the old mystiques of shepherding.

The provision of convenient handling yards which incorporate forcing pen, drafting race, foot-rot bath and dipper can make a major contribution in reducing the amount of time spent on routine tasks such as sorting, drenching and foot-rot control measures. Then there are devices such as drenching guns and cradles for holding sheep during foot-trimming operations which also make their contribution to a more efficient use of a shepherd's time.

There is also the part played by well-trained sheepdogs which despite the abuse and other more concrete things that are sometimes hurled at them, are indispensable for handling sheep. Britain is doubly fortunate in the quality of her sheepdogs. If there is a better sheepdog than a Border collie, which many will doubt, then unquestionably it is a Welsh collie, always provided it understands English (expletives included) if it finds itself on the other side of Offa's dyke.

Finally there is one area in sheep management where it does not pay to reduce the intensity of shepherding, and this is at lambing. We have seen the importance of a high rearing rate on gross margin per ewe and per hectare, and this is carried through, in a large measure, to profit because additional lambs do not involve very large additions to rearing costs. This is not just a question of adequate supervision at lambing but also of adequate protection of new-born lambs. They constitute the foundation of the only crop, apart from the meagre contribution from wool, that a sheep farmer harvests from his flock.

Chapter 10

DISEASE CONTROL IN THE EWE

As a corollary to the pressure for increased production has come a heightened awareness of the cost of losses due to ill-health, which should lead to a closer relationship between the sheep farmer and his veterinary surgeon. A resulting change in emphasis from disease treatment to prevention can have a marked impact on the average 10 per cent annual loss of sheep resources recently attributed to disease by economists at Exeter University. Unfortunately, a national Sheep Health Programme similar to that in pigs is not a current prospect, but at an individual or farmers' group level such programmes exist and are highly successful. These programmes are usually based on regular meetings at strategic times—such as prior to lambing and weaning—to examine the flock, discuss management plans, assess disease risks and devise or check preventive measures. It also gives an opportunity to review recent losses, and this emphasises one of the essential ingredients of such a programme, namely the careful recording of all disease incidents and deaths. An analysis of the cost of disease in lowland flocks showed that losses were lower in flocks with higher annual veterinary expenditure, the greater productivity more than offsetting the higher cost. With increased knowledge of disease causation and improved control methods, veterinary advice can only become more necessary to efficient sheep production.

Disease Causation
Disease may be classed as infectious or non-infectious and this to a large extent determines the control methods employed. Infectious diseases are due to biological organisms—the microscopic bacteria and viruses in one group, and the larger internal and external parasites in another—the

main factors influencing disease risk being the amount of infection to which the animal is exposed, and its own resistance to infection. While it may not be possible to avoid infection, management can be manipulated to reduce the infection challenge and to increase resistance. Non-infectious diseases are associated with simple nutritional deficiencies, or with the 'production diseases' in which a deficiency in dietary input is combined with the stress of production output. Again management methods can be adjusted to correct possible deficiencies and to minimise stress.

The diagnosis of disease is best left to the veterinary surgeon, who has access to specialised diagnostic laboratory services, and is best qualified to advise both on the treatment of specific infections and on the overall disease control policy on any particular farm. Nevertheless, the routine application of control measures to maintain the health of the flock depends on the farmer or shepherd, and the aim in this book is to provide an understanding of the major diseases and the principles on which control is based in the hope that this will improve the application of preventive medicine to sheep husbandry. Surveys of deaths from all causes indicate that the pre- and post-lambing period is the time of greatest risk, and in one study of hill sheep in Scotland 56 per cent of ewe deaths and 88 per cent of lamb deaths occurred at this time. Clearly this is where most attention should be paid and it is convenient to consider first the health and management of the ewe and then that of the lamb.

Ewe Health Problems

Failure of the ewe to conceive is apparently not often a problem, and there is some evidence that barren ewes have generally lost the developing embryo at an early undetected stage. This seems to be more common in young ewes, and good body condition and post-mating nutrition are important in minimising this loss. In late pregnancy and after lambing, problems can be considered under non-infectious and infectious causes.

NON-INFECTIOUS DISEASES

Given the demands on the ewe during winter, particularly where she is carrying twins or triplets, it is not surprising that nutrition and stress problems occur, and that the standard of management at this time is of crucial importance.

Pregnancy toxaemia or twin-lamb disease is still a common cause of death in ewes in late pregnancy, although improved nutritional standards have steadily reduced its incidence. It is characterised by nervous disorder accompanied by marked blood changes, in particular an abnormally low blood sugar level, and an increase in blood ketones. The latter substances are responsible for the smell of acetone (peardrops) on the breath of affected sheep, and also of cattle suffering from the related disease, ketosis. The association with multiple births suggests that this is a 'production disease' and, as with other diseases in this group, it is related to failure to match the input of a particular dietary component with the production demand. In this case it is a shortage of carbohydrate, the main source of energy, and the material from which the high glucose requirement of the foetus is met.

The factors which precipitate pregnancy toxaemia fall into two main types, the first being prolonged under-nutrition in the last third of pregnancy, and the second a sudden reduction in food intake in well-fed ewes combined with stress, such as severe weather conditions.

The under-nutrition type is seen in flocks receiving little or no supplementary feed during the winter and therefore on a declining plane of nutrition at a time when the demand is steadily increasing. Eventually, in late pregnancy, the glucose demand exceeds the supply and the ewe's blood glucose level falls. To counteract this, the body produces a hormone which stimulates the production of glucose from other sources, such as the fat reserves, and this is possibly the cause of the increase in ketones. In addition, the hormone is thought to conserve glucose by depressing the activity of other tissues. As the glucose shortage continues, the production of the hormone increases until its depressive effect interferes with the normal activity of the nervous system, and so the symptoms of pregnancy toxaemia develop, terminating in coma and death. In this type of case, treatment by supplying glucose or sources of glucose has little effect once symptoms appear, as the prolonged over-activity of the hormone cannot be reversed.

In pregnancy toxaemia associated with sudden deprivation of food coupled with stress in well-fed ewes, it is thought that the hormone stimulus following food shortage is exaggerated by the stress factors, and excessive hormone production has the same effect as before. This explains the common occurrence of the condition in heavily pregnant ewes in bad weather, especially in snow, which both reduces food intake and causes stress by increasing the body's requirements to maintain a normal temperature. This type responds much better to treatment in the early stages, for the brief stimulation of hormone production is more easily reversed before irreparable tissue damage has occurred.

A common factor in all cases is the failure of the dietary intake to meet the demands of late pregnancy, and this indicates the way in which the condition may be controlled. During the early stages of pregnancy the foetal demand is low and can be met on a low plane of nutrition, but in the last third of pregnancy the demand increases rapidly and the carbohydrate intake of the ewe must increase at a parallel rate. As a practical guide, the ewe should continue to gain weight, or at the very least maintain a steady weight, during the last six weeks of pregnancy, and supplementary feeding should be regulated to permit this. At the same time, flock management should ensure that no check in intake or abnormal stress occurs over this period, particularly in bad weather conditions.

Under intensive conditions aiming at higher lambing percentages, the risk of pregnancy toxaemia increases but should be offset by a higher standard of management and more accurate regulation of the quality and quantity of the feed intake. However, at high stocking densities and in housed sheep, care must be taken to allow adequate trough space to reduce competition and safeguard the intake of older ewes.

Specific mineral deficiency problems—lambing sickness and grass tetany—related to the demands for lamb growth in the uterus, and later for milk production are also common as might be expected.

Lambing sickness is a common disease of pregnant and lactating ewes, characterised by an abnormally low level of calcium in the blood and so resembling milk fever in the cow. The disease is sudden in onset and apparently not related to a simple dietary deficiency of calcium, so that the cause is not completely clear.

However, there are two important factors which give a clue as to the origin of the condition. The first is that the disease is restricted generally to late pregnancy and early lactation, when the demand for calcium is high. Secondly, it often follows a sudden change in environment or feeding, particularly the movement of hill ewes to lower pasture for lambing, and the risk appears to increase the closer the flock is to lambing. A combination of these two factors suggests that lambing sickness may result from additional stress at this critical period, affecting the animal's ability to mobilise its calcium resources to meet the high demand.

Two possible explanations have been suggested, one that there is a breakdown in the hormone mechanism which controls the mobilisation of calcium reserves from the bones, and the other that the stress causes a sudden fall in food intake, reducing the calcium available from the digestive tract. Neither has been conclusively confirmed, but the evidence available does indicate certain possible preventive measures.

Supplying additional calcium in the feed does not appear to be of value and prevention depends on the avoidance of additional stress at this time of high calcium demand. The essential precaution to avoid lambing sickness is the forward planning of the flock management to ensure a minimum of change and stress over this period of late pregnancy and early lactation. The movement of hill ewes should take place well before lambing if possible and, if the flock must be moved close to lambing, the change should be made slowly with intermediate rest periods.

Grass tetany or **hypomagnesaemia** is characterised by a low blood magnesium level. The incidence appears to be increasing and, as this seems to be associated with pasture improvement, the trend is likely to continue and it is a particular threat under intensive grassland management conditions.

The disease closely resembles grass tetany in cattle and the cause is probably the same in both classes of stock. It is most often seen in ewes with twin lambs and is particularly associated with a change from old grass to improved pasture, and with severe weather conditions, especially in early lactation.

An important breakthrough in the investigation of hypomagnesaemia came with the demonstration that only about 20 per cent of the magnesium in the diet was actually taken up by the animal, and that this percentage varied considerably with the type of diet. It was also found that, except in very young animals, the magnesium in the bones was fixed and could not be drawn upon. In fact, the body does not appear to be able to store magnesium in a usable form and so it relies almost entirely on its daily intake to meet its requirements for normal body activities and milk production.

If the daily balance of magnesium is of major importance, hypomagnesaemia might be expected to occur when the demand is high, and this is borne out by the association of the condition with peak lactation. Other associated factors are those which adversely affect either the intake or uptake of the element.

More accurate methods of analysis have shown that spring grass is low in magnesium, particularly in the young, rapidly-growing stage, and this deficiency is aggravated by a high potassium content, such as is produced by high fertiliser application, and this helps to explain the relationship between tetany and improved pastures. In addition, the uptake of magnesium in the gut appears to be lower from spring grass. Other elements, particularly calcium, may reduce the absorption of magnesium in the gut, and so liming might be expected to favour the occurrence of hypomagnesaemia. Individual variation in blood magnesium levels within a flock, and the variation in the incidence of tetany in animals with low blood levels, are not clearly understood, but may reflect the natural variation in efficiency of uptake and utilisation from animal to animal.

While many problems remain to be clarified, hypomagnesaemic tetany is basically a nutritional disease, associated with a low level of available magnesium in the diet, precipitated into a critical deficiency by any factor which affects the intake or uptake of this element, particularly when the demand is high. Prevention must, therefore, aim at maintaining a continuous intake of magnesium at a sufficiently high level to offset any of the factors which predispose to tetany. This is done by the provision of supplementary magnesium wherever the natural dietary source is likely to be low and in periods of peak demand, and in general these two situations tend to occur together in late winter and spring.

A number of methods of supplementation have been shown to be satisfactory. The use of magnesium-enriched concentrate ration, or a mineral mixture, or of calcined magnesite alone, is successful with sheep which are used to trough feeding.

Attention has also been paid to raising the dangerously low level of magnesium in tetany pasture to a safe level. The danger from heavily fertilised pastures has been stressed but, fortunately, it has been shown that the low magnesium levels in these swards can be raised to levels which effectively prevent tetany by the use of a magnesium-rich fertiliser, or top-dressing with magnesium salts in solid or liquid form. One heavy application of magnesium in this way has a persistent effect and has been shown to last for six grazing seasons.

INFECTIOUS DISEASES

Abortion: Of the infectious diseases, abortion is of course an ever present risk. The term abortion strictly refers to the production of a dead lamb before the normal period of pregnancy is completed. However, the important causes of abortion also result in a percentage of still-born and weakly live lambs which survive for only one or two days. There are a number of infectious diseases which can cause abortion as a side effect, particularly in association with a high temperature. It can also be induced by stress such as rough handling in late pregnancy. Occasional abortions therefore occur in most flocks, but when a number of cases occur—and above 2 per cent of the flock is suggested as a useful guideline—then the cause is likely to be one of the specific infections which only attack the developing lambs. These are, in order of importance, enzootic abortion or 'kebbing', caused by a virus; toxoplasmosis which is a protozoan infection; and two bacterial diseases, salmonellosis and vibriosis. Losses in these outbreaks can be as high as 30 per cent, and veterinary assistance should be sought as early as possible since diagnosis requires the laboratory examination of aborted material.

The distribution of these conditions varies, and low-ground flocks appear to be more commonly infected than hill flocks. This is probably because of the greater risk of transfer of infection at lambing time, infection being spread largely by contamination of pasture with organisms from the discharges and aborted material. Kebbing is probably the commonest cause, particularly in Scotland, but toxoplasmosis also appears to be quite widespread, although it has only been recognised relatively recently. Salmonellosis and vibriosis (the causal organism is now called Campylobacter) are much less common and may be declining in importance.

Infection of pregnant ewes with toxoplasmosis, vibriosis and salmonellosis causes them to abort straight away, but in kebbing the newly-infected ewes do not abort at that lambing but do so at the next pregnancy. Abortion is followed by immunity, so the common pattern is for a high level of abortion one year followed by a persistent lower level in susceptible young sheep and newly introduced animals.

Control is unsatisfactory except in the case of kebbing, for which a vaccine is available which gives a good level of protection. However, it will not prevent abortion in ewes which are already infected, and should therefore be given routinely to all ewes before their first mating. The other conditions present a particular problem, since there is a conflict between trying to limit the spread of abortion once it starts, and building up immunity in the flock by exposure to natural infection. In fact these organisms are highly contagious and it is almost impossible to prevent them from spreading, so that, in general, mixing to allow infection and subsequent immunity to occur is the most practical policy at the present time.

Border disease is a quite recently recognised disease which, although not causing abortion, can have a marked effect on lamb numbers. It is a virus condition and was first seen in the Welsh Border counties, hence the name, but is now spreading more widely. It results in a reduced lambing percentage and the birth of small lambs with long wavy hair in the birth coat, often showing nervous trembling. Such lambs (known as 'hairy shakers') usually die before weaning. Only one affected lamb is usually produced, the ewe fortunately being subsequently immune, since little is known about the transmission of the condition and its control.

Metritis: At lambing time constant attention is essential to care for the new-born lamb, but the care of the ewe is also important. A proportion of births will require assistance and the aim must be to obtain a living lamb with an absolute minimum of injury to the ewe. Damage will almost certainly be followed by infection, resulting in metritis or inflammation of the reproductive tract, and a common cause of death at this time. Veterinary advice regarding lambing hygiene and routine antibiotic cover for these cases can considerably reduce losses from this cause.

Mastitis is the main problem during lactation, not the chronic form so widespread in cattle, but as a more acute, sporadic condition particularly at weaning. It is responsible not only for an appreciable number of deaths but also for an estimated 5 per cent culling rate in the average flock, and therefore a significant economic problem. Symptoms are not easily seen until extensive damage is present in the udder so that treatment is not very useful, and prevention must be the aim. Fortunately a recent study has shown that long-acting antibiotics—the 'dry cow' preparations—are equally effective in sheep when administered at weaning, and this might well be a desirable routine in problem flocks, together with a change of pasture to reduce milk production.

Clearly the maintenance of the health of the ewe through pregnancy, lambing and lactation requires advance planning and an organised programme in relation to feeding, grazing management and the management of parturition, together with specific preventive measures against the major infectious diseases.

Chapter 11

HEALTH AND DISEASE IN THE YOUNG LAMB

LAMB SURVIVAL

Lamb health in the immediate post-lambing period is really a question of life or death since the bulk of lamb losses occur in the first fourteen days, and most of these on the first day. Stillbirths and injuries during difficult lambing account for a significant proportion of these early losses but the main cause is a combination of chilling and starvation. Cold stress is a problem of low birth weights—below 4 kg—and this is largely due to inadequate diet in late pregnancy. Twins show a higher mortality than singles for this reason, and the recent development of ultrasonic scanning to identify multiple pregnancies is an invaluable aid to the differential supplementary feeding of ewes according to the number of foetuses they are carrying. Chilling reduces the suckling drive and this reduces the energy available to the lamb, leading to low heat production and increased susceptibility to cold, setting up a cycle which rapidly ends in death. Shelter for the lambing ewes and continuous supervision to assist with lambing, prevent mis-mothering and ensure early suckling are therefore essential. In the Scottish study referred to earlier a loss of 8 per cent in one flock compared with an average of 15 per cent in nine others was attributed to overnight housing of lambing ewes and careful shepherding. The recognition that these losses are not due to infectious diseases is crucial, since it follows that they can only be prevented by improved management and not by improved drugs. The emphasis on what might be termed environmental losses does not, of course, imply that disease can be neglected, and a number of conditions are every-present risks to the lamb which must be controlled, largely by hygiene and preventive vaccination.

THE POST-LAMBING PERIOD

Coli bacillosis is a bacterial infection causing scours and death in lambs up to a week old, similar to calf scours and triggered off by a combination of overcrowding in unhygienic conditions plus digestive disturbance due to stress. It is a common sequel to chilling and poor suckling, resulting in a low intake of the first milk or colostrum which gives the lamb its first line of defence against infection. Its increasing occurrence is therefore probably a direct consequence of higher stocking rates and higher lambing percentage. The condition appears most often in the later lambs, particularly where the same lambing pens are used throughout. Antibiotics are effective in treatment, but the disease is better prevented by lambing small groups of ewes together and then changing the pens or lambing area to reduce the throughput in any one place.

'Watery mouth' is probably a related condition, also associated with a lack of colostrum, characterised by a wet muzzle due to regurgitation of stomach contents. Affected lambs show constipation, not diarrhoea, so antibiotics with laxatives are useful, and the same preventive measures apply.

Navel-ill or **joint-ill** is another disease associated with a high throughput in the lambing pens and particularly with a poor standard of hygiene, allowing an excessive build-up of normal skin bacteria. Through the unhealed navel of the new-born lamb, or through docking and castration wounds, these bacteria enter the bloodstream producing joint abscesses and consequently lameness. Control is relatively simple, again by improving lambing conditions, using clean pens which are changed frequently, and using properly sterilised instruments. The navel can be dressed routinely with a mild antiseptic.

Swayback. Mineral deficiency diseases would not be expected in the new-born lamb, but in fact a copper deficiency condition—swayback—is a well-recognised disease of the nervous system, restricted to lambs at birth or less commonly in the first few weeks of life. The explanation is that the deficiency is not in the diet of the lamb but in that of the pregnant ewe. In affected flocks a high incidence is associated with mild winter and warm spring conditions, i.e. with early lush growth of pasture. It has been restricted to particular areas such as south-east Scotland but is spreading in relation to marginal land improvement, the disease being commonly seen on limed and reseeded pasture, and therefore poses a distinct threat to increased productivity on marginal farms.

The symptoms are due to the degeneration of nerve cells in the brain and spinal cord, and the condition is associated with abnormally low levels of copper in the blood and tissues of affected lambs and their ewes. It is possible that copper-containing enzymes are necessary for the formation of nervous tissue, and a shortage of copper is generally recognised to be the basic cause of swayback.

In Australia, a similar condition occurs which is definitely associated with low copper levels in the pasture, and consequently a low copper intake in grazing stock, the critical pasture level being five parts per million. Modern analytical methods have shown that, in Britain, pasture copper levels in swayback areas are also dangerously close to or below five parts per million and, under conditions favouring rapid spring growth, this deficiency is accentuated. Furthermore, the availability of copper in the feed is markedly reduced by the effects of two other minerals, molybdenum and sulphur, and this is probably a key factor in inducing the deficiency condition. In general then, swayback appears to be basically a nutritional disease, associated with a lower copper availability in the pasture which is aggravated by the increase in sward production following improvement schemes or very favourable climatic conditions. Only a proportion of ewes with low blood copper levels give birth to swayback lambs, so that the relationship is not completely straightforward but is complicated by individual variation, which may reflect differences in metabolic efficiency from animal to animal.

Preventive measures are directed at ensuring an adequate intake of copper by the pregnant ewe. It can be supplied in a mineral mixture, or in a copper-enriched concentrate given to the ewes in the last six weeks of pregnancy. However, with this method there is little control over individual intake and consumption is erratic. Oral dosing with capsules containing copper needles, or the use of injectable copper salts provide more accurate and more reliable methods of suppmentation in a single dose. Control by any of these methods can be adjusted in accordance with the Ministry of Agriculture's forecast of likely severity. Top-dressing of pasture with copper compounds to correct the natural deficiency appears to be less satisfactory and is not recommended because of the risk of copper poisoning in grazing stock.

Copper toxicity. The increasing use of copper supplementation in foodstuffs has revealed the danger of chronic copper poisoning and a number of cases have been reported. It is now known that sheep are the most susceptible of farm livestock, because they can accumulate copper in the liver over long periods without symptoms appearing until a critical level is reached, after which a sudden breakdown occurs. Housed lambs appear to be particularly susceptible since hay/concentrate diets are often low in molybdenum and sulphur and therefore high in available copper. Sheep should not receive additional copper unless a deficiency has been confirmed, and a form which allows accurate dosage control, such as the injectable compounds, is to be preferred.

White muscle disease or stiff lamb disease is also a deficiency condition but seen in the slightly older lamb. It is associated with a lack of selenium, a mineral required in minute traces in the diet, and which can be highly toxic in excess. Until recently little has been known about its functions, but we now know that it acts in combination with vitamin E to prevent the accumulation of harmful waste products in the body which particularly attack muscle cells. This explains the deficiency symptoms of stiffness, lameness and even sudden collapse due to muscle degeneration, seen in lambs up to six months old, but particularly at 3–6 weeks of age. The condition appears to be on the increase in both calves and lambs possibly associated with the use of home-grown cereals in selenium-deficient areas, and this deficiency may also be responsible for some of the cases of ill-thrift in young stock. Low soil levels and therefore low intakes of selenium have recently been shown to be widespread in Britain, and 60–80 per cent of sheep in a national survey had apparently inadequate blood levels; but since only minute traces are required the economic importance of this finding is not yet established. Fortunately a diagnostic blood enzyme test is available, and veterinary advice should be sought to confirm a suspect flock rather than applying blanket medication. However, the availability of a soluble glass bolus, which is given by mouth, and releases copper, selenium and cobalt, over a twelve-month period, makes it possible to cover a range of related deficiencies in a simple way.

Coccidiosis is another condition seen in lambs at about 4–6 weeks of age, the extent of which is somewhat uncertain. The disease is caused not by bacteria but by a group of simple animal organisms—coccidia—and is best known as a major cause of acute diarrhoea and death in chickens. The problem in lambs is that while the infection has been quite conclusively associated with scouring and deaths, the organisms can be found in large numbers in healthy animals and high coccidial egg counts are common in dung samples from lambs. However, diarrhoea, often containing blood and associated with high dung counts at about a month old, should lead to suspicion of coccidiosis. The condition is becoming increasingly common, the number of outbreaks diagnosed by Veterinary Investigation Centres rising from 151 in 1975 to 502 in 1979. As with other infectious diseases, the increasing incidence is probably related to higher stocking densities with young lambs and increased opportunity for transmission, and while treatment with sulphonamides is effective, prevention by hygiene during lambing, when infection is probably passed from ewe to lamb, should be the first line of defence.

THE CLOSTRIDIAL DISEASES

In contrast to the previous conditions, which are only too common despite the general availability of control measures in most cases, there is a group of diseases caused by *Clostridia* bacteria which provides perhaps the major success story in sheep health. These diseases, ranging from lamb dysentery in lambs under two weeks old to tetanus in adult ewes, are potentially capable of causing enormous losses, particularly in lambs; yet due to the almost universal application of vaccination they are kept largely in check, and are therefore worth studying as an example of what can be achieved by careful planning and programming of health control measures.

The Clostridia can live normally in the soil, are widely distributed, and can infect animals at any time. They are frequently found in healthy sheep, usually in the intestines, and only cause trouble under particular circumstances, the special conditions which initiate disease being termed 'predisposing factors'. They cause disease largely by the production of extremely powerful poisons or toxins and tend to cause rapid death without characteristic symptoms, so treatment is virtually impossible and prevention is the only economically acceptable method of control. The problem of the clostridial diseases is aggravated by intensification, but fortunately the increased risk has been paralleled by advances in the efficiency of control by vaccination.

The intestinal diseases associated with clostridial infection are all due to one organism, *Clostridium welchii*. Three different strains of this organism are common in sheep, which produce different combinations of toxins and so cause three different clinical conditions. The first strain is designated *Cl. welchii* type B and causes the disease lamb dysentery.

Lamb dysentery is a condition restricted to lambs in the first two weeks of life. The disease risk is widespread but the incidence varies considerably from year to year, both in the number of lambs affected on each farm and the number of farms involved. It is thought that the bacteria are commonly picked up from the soil or from the teats of the ewe and that, under certain conditions, they multiply excessively in the intestine, producing toxins which are absorbed causing severe inflammation and death. The predisposing factors which trigger off this disease are not known.

Entero-toxaemia: The other two strains of *Cl. welchii*, type C and type D, also cause disease by excessive multiplication and toxin production in the intestine, but affect animals over two weeks old. This condition is termed entero-toxaemia and it is of considerable importance to identify which strain is responsible for any one outbreak. Type C organisms are relatively uncommon, and only appear to cause disease in adult animals, sometimes termed 'struck'. By contrast, type D organisms are widespread and can cause disease in sheep of any age, excepting lambs under one to two weeks old. Type D disease in sheep under one year old is often termed pulpy kidney disease, because a softening of the kidneys is a common post-mortem finding.

Entero-toxaemia was formerly one of the commonest and most important killing diseases, particularly from an economic standpoint, as it tends to strike the best thriving animals on the best grazing. It is caused by the rapid multiplication of the organisms in the intestine and the resulting production of large quantities of lethal toxins. The most important predisposing factor appears to be the appearance of poorly digested food in the intestine, which offers a more nutritious diet for the bacteria, allowing a more rapid rate of growth and multiplication. This often follows a sudden improvement in the nutrition of the sheep with which it cannot immediately cope, and so entero-toxaemia is generally associated with dietary changes.

Braxy, caused by *Clostridium septicum*, is also associated with the alimentary tract, but in this case the organism is found in the fourth stomach. It appears to be triggered off by feeding on frosted material and so tends to occur mainly in young hill sheep in their first exposure to winter conditions.

The remaining clostridial diseases form a somewhat separate group, associated not with the alimentary tract but with other tissues, and set off by much simpler predisposing factors, namely any injury to the tissue. The group comprises gas gangrene, tetanus and black disease, and the organisms are commonly found in the soil and in the body, so that wound infection can occur easily.

Gas gangrene is due to wound infection with *Clostridium septicum* or with *Clostridium chauvoei*, the cause of black-quarter in cattle. Infection is typically associated with the wounds caused during shearing, docking and castration, and with internal damage caused by a difficult or assisted lambing. The condition is, therefore, a major threat in spring and early summer in ewes and lambs.

Tetanus also arises by the infection of external wounds, such as docking and castration, and with umbilical infection, but womb infection is uncommon, so that the risk is generally restricted to young lambs. The infecting organism is *Clostridium tetani* and the disease has a different appearance, because the toxin affects not muscle tissue but the nervous tissue, making the animals hypersensitive to any sort of stimulation.

Black disease is the result of infection in an internal organ, the liver, the organism *Clostridium oedematiens* being found, apparently harmless, in the normal liver, but multiplying excessively if the liver tissue is damaged. The condition is, therefore, closely associated with liver damage from heavy fluke infection in the autumn and, in Wales, is considered to be responsible for half the deaths which occur during outbreaks of acute liver fluke disease.

THE CONTROL OF CLOSTRIDIAL INFECTIONS

The widespread distribution of clostridial organisms in the soil and the animal body makes the control of this group of diseases by normal husbandry methods virtually impossible. Attention to hygiene in castrating and docking and in assisted lambing, the use of clean lambing pens changed at frequent intervals, and fluke control measures, will obviously reduce losses but will not give complete control.

It must be recognised that intensification increases the risk of clostridial disease, but at the same time the demand for increased efficiency of production makes it essential to prevent losses from this cause, and for these reasons planned vaccination programmes are advocated to give protection from the group as a whole. It is, however, essential that a planned programme should be conscientiously adhered to, for as the organisms persist indefinitely in the soil they cannot be eradicated, so that the risk of infection remains and immunity lasts only as long as vaccination is maintained (Plate 6).

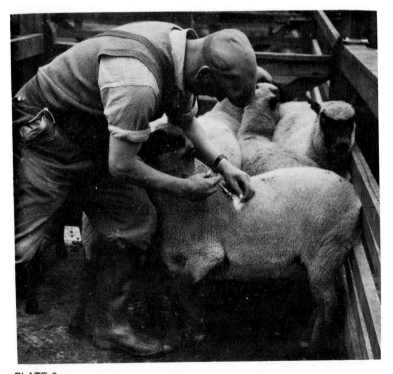

PLATE 6
Instructions regarding site and method of vaccination should be carefully followed.

Photo University of Newcastle

In the case of the Clostridia, damage is caused by toxins and these stimulate the production in the body of antibodies known as antitoxins which neutralise the toxins. In a normal flare-up of clostridial disease, antitoxin production is too slow to prevent the lethal action of the toxins and death occurs rapidly. However, if antitoxin production has previously been stimulated by vaccination, then in response to a flare-up a very rapid build-up of antitoxins occurs which renders the toxins harmless. In the clostridial vaccines the stimulating substance consists of toxin which has been rendered inactive by chemical means.

One other aspect of immunity is very important, and that is the ability of the mother to concentrate antibodies and secrete them in the colostrum or first milk. These antibodies are absorbed by the suckling young animal and persist for a few weeks, giving immediate temporary immunity. Vaccination in pregnancy can be used to raise the maternal antibody level and so raise the concentration passed to the offspring, producing a high level of immunity for the dangerous early weeks of life. Subsequently, the young animal can be vaccinated to produce its own long-term protection.

It is possible to provide immediate temporary protection in a susceptible animal by transferring antibodies from an immune animal in the form of a concentrated blood extract known as antiserum.

The important difference between the protection conferred by vaccination and that produced by antiserum is that vaccination stimulates the animal's own antibody production, which takes some weeks to become effective but lasts for months or even years, whereas antiserum provides ready-made antibodies giving immediate protection but lasting only a few weeks. For long-term protection, vaccination at relatively long intervals is obviously the method of choice, but for immediate short-term protection of susceptible animals in the presence of a high risk of infection, antiserum is of great value.

The aim of a planned vaccination programme is the maintenance of an adequate antibody response to overcome any challenge caused by a flare-up of bacterial activity, and this is required right from birth. However, the ability to produce antibodies declines steadily and requires regular stimulation by revaccination to persist at a high level.

The annual programme starts with the injection of the pregnant ewe. This injection raises the immunity of the ewe herself and also results in a high antibody concentration in the colostrum, which transmits an effective level of protection to the new-born lamb. At this stage, protection is required against gas gangrene and entero-toxaemia type D for both the ewe and lamb, against lamb dysentery and tetanus for the lamb alone, and possibly against entero-toxaemia type C for the ewe.

This protects the lambs for up to 10–12 weeks, after which lamb dysentery, gangrene and tetanus are no longer a problem, but entero-toxaemia type D still is and the temporary immunity should be replaced by vaccination at 8–10 weeks. Lambs retained for breeding should, at this stage, begin their full protection programme against all those clostridial infections likely to be encountered, either during adult life or by their future offspring, but fattening lambs need protection only against entero-toxaemia type D. Where braxy or black disease occur, additional protection against these will be required in the autumn.

The advent of multiple vaccines has greatly reduced the time and labour required to provide full protection against the Clostridia, and a number of vaccine combinations are available to provide patterns of protection, up to the eight-in-one preparations which protect against the whole group.

Chapter 12

GENERAL FLOCK HEALTH PROBLEMS

Apart from those conditions particularly affecting the ewe and lamb, there are individual diseases which represent serious actual or potential threats to efficient production, and in general intensification favours their occurrence. To a considerable extent this is offset by advances in veterinary preventive medicine, particularly vaccine development, but intensificaton also offers possibilities for improvement in management and nutrition. Closer control of the flock means better control of the environment, often the main source of trouble, and permits higher standards of hygiene. Increased productivity also justifies increased expenditure on preventive measures and on such things as handling facilities. A number of examples are worth considering to illustrate different approaches to disease control, and in some cases to draw attention to the problems which still exist.

FOOT-ROT

This disease is an excellent example of a condition in which lasting economic benefits can be obtained by a short-term increase in effort and expenditure on comprehensive control measures. It is still a major source of economic loss because of its widespread occurrence and severe effect on thrift, and in fact is specifically mentioned in the Code of Welfare for sheep, which recommends that stockmen should be experienced in its control. Yet it has been shown to be eradicable on a farm scale and control is a relatively straightforward matter of good husbandry. A better understanding seems necessary among shepherds of the principles on which control is based, and particularly of the economic cost of foot-rot in lowered productivity.

The condition itself is well recognised and, while it is not the only cause of lameness, it can be differentiated from other possible causes. The greatest contribution to foot-rot control was the demonstration that it is caused by a specific infection which spreads from foot to foot via contaminated pasture. Furthermore, the organism concerned lives only in diseased feet and will not survive for longer than about a week in the soil. It is maintained in a flock in the feet of chronic carriers, in which small pockets of infection heal over under dry conditions and partial recovery results, followed by a breakdown when damp conditions return. Under wet conditions which favour the entry of the organisms into the foot, the resulting pasture contamination spreads the infection widely, and therefore severe oubreaks are associated with soft conditions underfoot. However, wet conditions will not produce foot-rot in the absence of the organism, and it can be eradicated by eliminating this organism from the flock and then preventing its reintroduction.

Before embarking on an eradication programme, it is necessary to assess the economic benefit to be obtained and the feasibility of the required control measures under normal farming conditions. The advantages of freedom from foot-rot have been convincingly demonstrated in this country in terms of the performance of affected and unaffected sheep, and the practical possibility of eradication has been confirmed in a variety of commercial flocks.

Much of this work was carried out some years ago at the Central Veterinary Laboratory, Weybridge, where in a typical trial an average liveweight difference of 3·3 kg was recorded between healthy and diseased sheep, which was reduced to 1·6 kg five weeks after treatment. However, the fact that such a difference persisted several weeks after final clearance of the disease emphasises that prevention is better than cure. While this result shows the value of treatment and eradication in terms of meat production, it may obviously be interpreted on a wider basis as indicatiing the degree of interference with thrift due to the disease and a justification for its control.

Control programme

The control programme is based on three principles. First, the organism concerned is found only in the feet of sheep and goats, so that no other class of stock is usually involved apart from the sheep flock. Secondly, it can be cleared completely from almost all affected feet by treatment with a range of preparations, as long as the foot is properly prepared to expose all the diseased tissue, and only the most chronic cases have to be disposed of, Thirdly, the organism does not survive longer than one to two weeks on contaminated pasture and pastures can be completely cleared by a two-week rest from sheep. A typical routine is as follows:

1. The examinaton and trimming of every foot and the separation of all suspect cases, including any showing moist inflamed areas which may be early cases incubating the disease, and any with very badly-shaped feet which may be symptomless chronic carriers. The healthy sheep are then run through a foot-bath (Plate 7) containing 5–10 per cent formalin, immersing the feet for a full minute, stood on a clean floor for an hour to allow the feet to dry, and turned on to a pasture which has not carried sheep for two weeks. This group should subsequently be watched carefully for any breakdown in control.

2. Treatment of affected feet. First, careful paring away of. all the separated horn to expose the underlying infected tissue is essential and must be carried out thoroughly but without injury to the living tissue below. Lack of attention at this stage is thought to be the main cause of breakdowns following treatment (Plate 8).

The prepared foot is dressed with one of a variety of preparations, such as 10 per cent formalin and broad-spectrum antibiotics, but copper sulphate seems less satisfactory. Formalin has a marked price advantage over the antibiotics but is rather unpleasant to handle, and deciding the relative merits of cost versus convenience is an individual matter.

3. The treated group is then kept separate from the main flock for re-examination and retreatment at weekly intervals for three or four weeks. All the recovered animals are run through the footbath and added to the clean flock, or maintained on clean ground as a separate unit for a further month

PLATE 7
A sloping-sided foot-rot bath will accommodate sheep of different sizes.

Photo University of Newcastle

to guard against breakdowns. Any persistent cases at this stage should be sent for slaughter as these would be likely to remain carriers indefinitely.

Modifications can be made to suit different conditions as long as the basic requirements of careful inspection, thorough preparation and treatment, and clean pasture are observed. For example, where the number of affected animals is very small it may be possible to dispose of them immediately, and subsequently all that is necessary is a close watch for any new cases in the clean flock.

The programme is best undertaken during a dry spell in

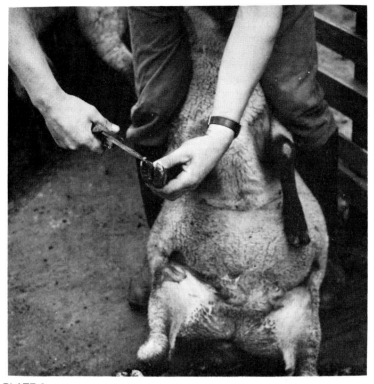

PLATE 8
Careful examination and trimming of feet are essential in foot-rot control.

Photo University of Newcastle

summer when the incidence is low and there is a minimum
risk of animals being in the early stages of the condition
without showing obvious signs. The healthy flock should be
carefully watched, and should be re-examined and run
through the footbath at least once during the programme.
Finally, all sheep subsequently introduced should be
examined and, if necessary, treated and isolated before they
are allowed to join the flock.

This procedure has been shown repeatedly to be effective
in practice, and although a considerable amount of work is
involved over a short period, it is much less than that
expended in repeated annual treatment, with the advantage

of a more permanent result and greater economic benefit. A vaccine against foot-rot is now available and is proving effective against this specific infection, used either to speed up eradication by preventing relapses and new cases, or for routine control. The latter necessitates vaccination every six months but is highly recommended. However, it must be emphasised that a good deal of the lameness encountered in flocks is due not to true foot-rot but to a variety of similar conditions, particularly 'scald', the actual causes of which are not properly understood. These conditions, which will not be prevented by vaccination, are much less crippling and only flare up at irregular intervals; but they do still make it necessary to maintain a good standard of foot health by the use of the footbath and the trimming of overgrown feet.

ORF

Contagious pustular dermatitis (orf) is a virus infection which specifically attacks the skin, forming scabby patches, particularly around the mouth and around the feet, thus being another cause of lameness. It may also spread to the soft tissue lining the mouth and the genital organs. The infection is most active in spring and summer, and as older sheep develop a fairly strong immunity, it is usually restricted to animals under a year old, often spreading rapidly through this age group each year in affected flocks.

The severity of the condition depends largely on the age of the lamb affected, being commoner and less severe in older weaned lambs. It is less common in suckling lambs, but in this age group the skin is much more extensively affected, while occasionally the immunity of the ewes is broken down and the teats and udder become involved.

The infection appears to be carried by recovered animals and the dried scabs remain infective for many months, so that it is extremely difficult to eliminate the condition. However, a suitable vaccine has been developed using living virus, which requires careful handling and can only be stored for a short period. Vaccination of lambs is carried out in the early summer, but not in the first month of life because in very young lambs the living virus may produce a severe reaction rather than immunity. Although this condition is not normally fatal, it does cause a serious check at a time when maximum growth rate is required and warrants this simple precautionary measure in areas where it is common.

PNEUMONIA

Acute pneumonia causes serious losses in sheep and is likely to become increasingly important, as some of the factors which appear to precipitate the disease are more common under intensive systems of management.

The disease occurs in all ages of sheep, particularly in spring and autumn, and is associated with bacteria of the *Pasteurella* group. These organisms are common in the nasal passages of normal sheep and spread to the lungs to set up inflammation under certain conditions, which are not clearly identified, but often involve bad weather or the close confinement of animals for feeding, housing and transportation. In housed lambs acute pneumonia has been associated with feeding high levels of barley without adequate roughage. Heavy lungworm infection may also cause pneumonia, particularly in hoggs in autumn, and is usually complicated by secondary *Pasteurella* infection.

Treatment is not very satisfactory and, as the organisms are found in normal sheep, the infection cannot be eliminated although measures can be taken to prevent a flare-up. Vaccination gives useful protection and should be combined with careful attention to predisposing factors, such as management changes, to minimise the risk of an outbreak. For example, the condition is common in housed sheep, but it is unlikely to occur if the animals are vaccinated before being housed and are given adequate space and ventilation. The importance of predisposing factors in acute pneumonia emphasises the necessity for higher standards of management as intensification increases.

In addition to acute pneumonia, there are a number of other forms of the disease which are less well recognised and even less clearly understood. Sporadic cases of pneumonia occur in housed animals, particularly in their first winter, and chronic pneumonia is also seen, though much less commonly, associated with progressive cancerous changes in the lung tissue. While these conditions are not yet very significant in this country, a similar chronic pneumonia in housed sheep in Iceland did become a very serious problem called Maedi-Visna, and the specific virus responsible has recently been isolated from sheep in this country, probably as a result of importations from Europe. A Maedi-Visna Accredited Flock Scheme is in operation to maintain flocks free from this condition.

SCRAPIE

This is an example of a disease for which no effective treatment exists, and yet which can be eliminated from a flock by management methods. Its economic importance is difficult to assess, but in general it is probably not a major problem, although it has affected pedigree exports and can cause serious disruption of individual breeding flocks.

It is characteristically a chronic wasting disease associated with changes in the brain and nervous system. Usually only one or two cases are seen each year in cross-bred flocks where the lambs are slaughtered, but in breeding flocks the incidence often rises rapidly and drastic measures have to be taken to control it.

The main problem has been to determine the method of transmission. There is strong evidence of close family relationships in affected animals and this led to the suggestion that the disease is transmitted through breeding, as attempts to infect animals by contact had been unsuccessful. However, there is increasing evidence to show that it will spread by contact to normal sheep exposed to heavy contamination, and it has been transmitted by feeding infected tissue. The rate of spread of infection in pure-bred flocks is much higher than would be expected with contact transmission alone, and in practice it is likely that a combination of both contact and congenital transmission is involved, while the genetic pattern of the animal may be of importance in determining susceptibility.

In this country, where the condition is widespread and eradication probably impossible, most attention should be given to controlling transmission through breeding by eliminating the affected animal together with its parents and offspring, particularly in the female line, as the ewe appears to be more important in transmission than the ram. This requires careful recording which is difficult but has been shown to be highly effective. Control therefore necessitates a flock recording scheme involving ear-tagging all ewes and rams, and all lambs which may be kept for breeding, in order to relate parents and progeny.

PINING

While swayback in young lambs is the most obvious mineral deficiency problem in sheep, the condition known as pining is perhaps a more typical deficiency disease, being associated with chronic unthriftiness rather than specific symptoms. As the name suggests, it is a wasting disease of ruminants, sheep being more susceptible than cattle, and is due to a shortage of vitamin B_{12}. This vitamin does not occur naturally but is produced by a variety of bacteria, including those found in the ruminant stomach. However, as cobalt is part of the vitamin B_{12} molecule, the vitamin cannot be synthesised unless an adequate supply of cobalt is available and the vitamin B_{12} shortage responsible for pining is due to a lack of cobalt in the diet.

The condition is associated with a simple deficiency of the element in the pasture, due either to a deficiency in the soil or to the chemical composition of the soil making the cobalt unavailable to the plant. This second type of deficiency is seen on soils with a very high calcium content, as found in Australia and parts of Northern Ireland. The effects are most marked in the spring, when active growth depresses the pasture cobalt level still further, and in young animals in which the demand for vitamin B_{12} is relatively high. Wherever the condition occurs, even in a mild form, it is of considerable economic importance in limiting production, for any attempt to increase productivity from such pastures will accentuate the deficiency.

The condition can be tackled in one of two ways, either by the direct administration of cobalt to the animal or by the correction of the deficiency in the herbage.

The simplest method is the provision of mineral mixtures containing cobalt, and this is satisfactory where the animals are used to supplementary feed. Individual dosage ensures that each animal receives an adequate amount but experience has shown that, for best results, the flock should be dosed at weekly intervals. Regular dosing is, however, only possible with small flocks and, for large groups, the cobalt 'bullet' has been developed. This is a large, heavy pellet containing cobalt which is administered with a balling gun and passes into the first or second stomach, where it remains and slowly dissolves, giving a steady release of cobalt into the stomach for bacterial utilisation. Theoretically, the pellets contain enough cobalt to last for several years but, in practice, their effective life is probably not more than twelve months. The soluble glass bolus, referred to earlier for supplementation with copper and selenium, also contains cobalt and provides long-term supplementation of all three elements in a very convenient form.

The correction of the pasture deficiency is, in general, a more satisfactory method of prevention and is simplified by the safety of cobalt compounds, the small quantity required, and the long duration of effect. In Britain, the application of 2 kg of cobalt sulphate per hectare every four or five years is considered sufficient, but experience in Ireland has shown it to be less satisfactory on highly-calcareous sandy soils, where an excess of calcium carbonate limits the uptake of cobalt by the herbage.

Chapter 13

PARASITIC DISEASES

There are two groups of parasites, external parasites living in or on the skin, and internal parasites living mainly in the digestive system and the lungs, the chief problem in both groups being the transfer from one host animal to another. In the course of evolution, special modifications have developed to ensure efficient transfer in natural conditions where the host animals are freely dispersed. Under domestication the movement of stock is limited and the density increased, resulting in easier transfer and abnormally successful host infection, so that parasitism has become possibly the most serious cause of production loss.

Parasitic diseases are common throughout the world, and until recent years the aim of control has simply been to check the build-up of parasite numbers and prevent obvious disease. However, with the development of highly effective drugs, and a much better understanding of parasite behaviour, we can and must design control programmes to suppress parasitism to a point where there is not only freedom from disease outbreaks but also no signficant interference with productivity. Parasite control is therefore in a period of rapid development made possible by chemical and biological advances, and to capitalise on this requires the maximum co-operation between veterinarian, agricultural adviser and farmer.

EXTERNAL PARASITES

External parasites belong to a biological group which includes the insects, mites and ticks, and show considerable similarities in structure, behaviour and susceptibility to chemicals. This means they can be attacked as a group and so we look for

167

preparations which, while being highly effective against the most important types, have a wide enough range of activity to control the whole group with a relatively simple programme. In general, these parasites are only accessible to chemical attack while on the sheep, and the ease with which they can be controlled largely depends on the length of time spent on the host. The easiest to control are those which spend their whole life on the sheep, namely the mites, lice and keds, and which cannot survive off the host for more than one or two weeks. Effective control of this group is obtained by a single whole-body treatment with a preparation which is lethal to adults and young stages and persistent enough to kill the larvae as they hatch from the resistant eggs, or by repeated use of a compound which only kills the adults. This was the basis of the double-dipping and later single-dipping regulations for the control of scab mite, and the compulsory treatment of all suspected cases led to the eradication of the condition once the highly effective DDT and BHC preparations became available.

Mites. The only important mite in sheep is the one causing scab. Less than pinhead size, it resembles a miniature tick and lives on the skin where the wool is thickest, sucking tissue fluids. Excess fluid dries and thickens, detaching the wool and leaving the charactistic bare scabby patches. Eggs, young stages and adults are all confined to the sheep, so that it is well controlled by dipping, and since it is seasonally active in winter an autumn dip is the most effective. Spraying unfortunately is less reliable and therefore not acceptable.

The condition was reintroduced in 1973 presumably by imported sheep, and partial control by tracing and dipping in-contact animals failed to stop its spread. Compulsory dipping on a national basis is eliminating this parasite again, but progress is slower than expected due to failure to dip and to dip thoroughly, and it is in the interests of all flock owners to comply conscientiously with the regulations.

Lice and keds were effectively controlled by the scab treatment and have subsequently been controlled as a side-effect of flystrike prevention. However, lice resistant to the common insecticides appeared in north-west England, rapidly resulting in serious infestations, and while the changeover to organo-phosphorus compounds for fly control quickly restored the position, this incident emphasises the necessity for continuous control. One benefit of the reintroduction of compulsory dipping for scab has, of course, been a reinforcement of lice and ked control.

The blowfly completes only one fairly short stage of its development on the sheep and so is open to attack for a much more limited time. The adult fly lives for about a month and lays batches of eggs on a number of sheep. Larvae hatch rapidly and feed on the skin and tissues for a period of days, then drop off and form a resting stage or pupa in the soil from which the adult fly emerges after a few weeks. A number of generations of flies develop over the activity season from May to September, the final generation overwintering in the soil to emerge the following spring.

The problems of blowfly control are the relatively short susceptible period on the host, the repeated infestations which occur throughout the long activity season, and the rapidity with which severe damage develops. A preparation is required which is rapidly effective against the larvae, and which gives persistent protection. Only in recent years have such chemicals become available. The earliest arsenical compounds, which persisted for only two or three weeks, were replaced by DDT and BHC with an effective duration of four to seven weeks, and then by dieldrin which persists for ten to twenty weeks, giving effective control with one seasonal treatment. However, in 1965 dieldrin was withdrawn because of the dangers of persistent tissue residues, and its place was taken by organo-phosphorus and carbamate compounds. These have been generally satisfactory, but tend to give slightly shorter protection than dieldrin.

The choice of the method of application of the insecticide is largely influenced by farm conditions and lies between spraying and dipping. The high incidence of shoulder strike in Britain makes whole-body protection necessary and so the Australian practice of using high-pressure jets on the hindquarters solely to control crutch strike is not favoured. In general, the plunge dip is simpler to operate and gives more effective protection, but sprays have the advantage of mobility, speed of operation, economy of insecticide and less injury risk. Where no special factors operate to favour spraying, the dip is probably still the best method of application, and in any case must now be available for scab control.

Simple hygiene is still important, and regular crutching, treatment of scours and dressing of wounds are essential, for soiled and injured areas increase the attractiveness of the sheep to flies and hasten the breakdown of protection. Early signs of strike, such as vigorous tail-wagging and moist discoloured patches of wool, must be watched for as indicators of failing protection, for the duration of effectiveness varies, particularly with the intensity of fly attack. This is particularly important with the organo-phosphorus compounds, whose protective effect tends to terminate abruptly. Treatment of struck sheep is best carried out with a mild antiseptic cream containing BHC to kill the larvae; strong antiseptics should be avoided.

In northern England and southern Scotland, a more specific fly problem has apparently increased in frequency and severity in recent years, namely 'Headfly'. This particular fly, which resembles the ordinary house fly, feeds on the fluids from the eyes, nose and mouth of the sheep. As a result of the irritation the animals rub their heads against fences, walls and other rough surfaces causing skin wounds which are then even more attractive to the flies, consequently increasing the damage and spreading the affected area. The result is a broken head and a disturbed unthrifty sheep, and an obvious economic loss especially in the sale of ram lambs.

Its increased occurrence has been attributed to the withdrawal of dieldrin, and while this is not proven, it is clear that dipping is not particularly effective in prevention, although this may really be due to lack of wool on the head and consequently less retention of the chemical. Also it is now known that the larval and pupal stages occur in soil and are not susceptible to control measures. Repellents and even head caps have been used to protect infected animals, but the new synthetic pyrethroid compounds applied as pour-ons show the most promise for effective control of this parasite.

The tick spends relatively even less of its life cycle on the sheep and only attaches for a period of about a week to take a blood meal each year during its three-year cycle of development. The rest of the year is spent in the pasture, where it requires a high moisture level for prolonged survival and so is restricted to rough grazing, woodland and other areas with a thick moist layer of vegetation. The main activity period occurs in spring, usually April–May, with a second period involving smaller numbers in the autumn. The majority of ticks are, therefore, only accessible on the host over this two-month period, during which individuals attach for only one week and a large turnover in tick burden occurs.

Tick control requires a preparation highly effective against all the developmental stages, and protective for at least two months. This might appear easier than the more prolonged effect required for blowfly control but, in practice, tick control is the more difficult of the two.

BHC and the organo-phosphorus compounds are highly effective, but the former persists only two to three weeks, and the latter up to six weeks. Furthermore, the activity period coincides with lambing in the main tick areas of Scotland and northern England, so that treatment must be carried out before lambing and cannot be repeated without considerable difficulty. Full protection for the whole period of tick attack is therefore unlikely. In addition, tick control in the lambs presents a further problem, for dipping is difficult at this age and they usually receive much less effective protection, such as the application of derris powder or a BHC cream to the most vulnerable areas. The pour-on synthetic pyrethroids will overcome many of the difficulties of tick control, and are clearly preferable for lamb protection, but unfortunately they cannot be used on lambs under a week old. Finally, as the host is merely the source of a blood meal, the tick can feed on a variety of animals and control on the sheep will not eliminate the parasite, while relatively small numbers are dangerous as transmitters of disease.

Other measures to reduce the tick population are of comparatively little significance. Pasture improvement produces conditions less suitable for tick survival, but its possibilities are limited in hilly areas. The elimination of alternative hosts is difficult, and the application of chemicals to kill ticks on pasture is not yet practicable. Tick control, therefore, remains a considerable problem and its solution would greatly improve hill farm production. With this in mind there will be great interest in the application of a new injectable group of drugs—the avermectins—which have a dual activity against external parasites and roundworms, and might prove applicable to tick control.

Tick-borne diseases
The economic importance of the tick as a sheep parasite lies not so much in the damage caused by the tick itself over the relatively short period of attachment to the host, but in a group of three conditions, two of which, louping-ill and tick-borne fever, are directly transmitted by the tick, while the third, tick pyaemia, only occurs in association with tick infestation. These diseases are limited by the distribution of the tick, but unfortunately this coincides with the distribution of a considerable proportion of the sheep population, and they represent a considerable stumbling block to increased productivity from hill farms.

As they are all associated with tick infestation, infection is largely confined to the main season of tick activity in spring and early summer, coinciding with the lambing season and resulting in a high incidence in the lamb crop, following which immunity develops. Older sheep are only affected if they escaped infection in their first season or have been newly introduced to an infected area.

Tick-borne fever is the most widespread of the three conditions and also shows the highest infection rates within flocks. The organisms are transmitted to the sheep when infected ticks start to feed and remain in the blood of affected sheep for some weeks, so that these animals act as a reservoir to infect a high percentage of the next generation of ticks, and this results in a high level of infection in each year's lamb crop, followed by immunity. The disease is usually very mild in lambs and few deaths occur. In older animals it occurs only in sheep brought in from a tick-free area, and in these it is more severe and can result in abortion in pregnant ewes. This disease does not cause much direct economic loss but the presence of this widespread infection apparently increases the severity of louping-ill and tick pyaemia.

Louping-ill is probably the most important of the group economically, causing an appreciable death rate in lambs and yearlings. In many animals a mild blood infection develops, followed by recovery and subsequent immunity, but in a proportion of lambs, and increasingly in older sheep, the infection passes from the blood to the nervous system and death results. As the blood infection lasts only a few days relatively few feeding ticks become infected, and therefore the rate of infection in the lamb crop is much lower than with tick-borne fever. Those lambs which escape infection in their first season remain susceptible and are likely to become infected as yearlings or even later, and so louping-ill is not restricted to the lamb crop.

Tick pyaemia is entirely restricted to the lamb crop and affects a high proportion of flocks in tick areas. The organism concerned does not infect the tick but is found on the skin of the lamb and is thought to enter the body accidentally through the tick-bite, resulting in widespread abscess formation especially in the limbs, causing a characteristic lameness known as 'cripples'. Lambs suffering from tick-borne fever have been shown to be a hundred times more susceptible to tick pyaemia than normal lambs, suggesting that the simultaneous presence of tick-borne fever may be an important predisposing factor in natural outbreaks of the disease. This would explain why tick pyaemia is limited to the lamb crop, being associated with high incidence of tick-borne fever in this age group.

Complete control depends on total tick eradication and this cannot be achieved with the level of tick control which is normally obtained. The difficulties involved in controlling ticks have already been discussed, but of particular importance in relation to these diseases is the fact that the highly-susceptible lambs may pick up ticks within a few hours of birth, so that while the application of insecticidal creams and other measures will reduce tick numbers, it will not fully protect the lambs.

A vaccine against louping-ill is available which is effective over a relatively short period. Administered to susceptible older sheep in spring just before tick activity begins, it provides a temporary immunity which will become permanent if active infection occurs as a result of the bite of an infected tick during this time. It may also be used to protect lambs against an autumn infection where two periods of tick activity occur, but is less successful in very young lambs in spring.

No methods of prevention are available against the other two conditions and the only one of the three which responds to treatment is tick pyaemia, in which early treatment with penicillin considerably hastens recovery.

INTERNAL PARASITES

Whereas the external parasites form a reasonably uniform group, the internal parasites differ considerably in position, behaviour and drug susceptibility, so that control methods do not follow one basic pattern. The three broad classes are the tapeworms, the flatworms represented by the liver fluke, and the roundworms of the digestive system and the lungs.

The tapeworms require two hosts, the first harbouring the adult tapeworm in the gut, where it lays eggs which pass out in the dung of the host. These eggs are picked up in feeding by the second host, in which a young stage hatches and migrates into the tissues to form a bag-like 'bladderworm'. Further development to the adult only occurs if the bladderworm is eaten by the first host.

The sheep is host to only one adult tapeworm, the 'milk-worm' which is largely restricted to the suckling lamb. The second host is a small pasture mite and lambs become infected by taking in these mites in grazing. This tapeworm is generally of little economic importance, but there are reports of unthriftiness, scouring and even death, apparently due to heavy infections, so that it cannot be completely ignored. No attack on the pasture mite is possible and routine treatment of lambs is generally unnecessary, the infection being thrown off by the time they are weaned. However, where trouble occurs highly effective remedies are now available, and some of the newest broad-spectrum anthelmintics are very active against tapeworms as well as roundworms.

The sheep can also harbour a number of bladderworms, the eggs being picked up on grazing contaminated with the droppings of dogs and foxes, which carry the adult tapeworms, and are themselves infected in turn by feeding on sheep carcasses. The bladderworms are thin, fluid-filled sacs, usually up to 50–80 mm in diameter, and the commonest species is found among the intestines where it does little damage. However, the bladderworm of a second species tends to develop in the brain causing the condition known as 'gid', which is characterised by abnormal turning movements. 'Gid' is not uncommon and is of some economic importance, but it is also a public health problem since the bladderworm can develop in man. This is relatively rare, but another bladderworm, known as a hydatid cyst and also originating from a dog/fox tapeworm, is much more infective to people causing serious damage due to its relatively large size. Farming families are obviously particularly at risk, and in a survey carried out in Wales, where the highest infection rates are found, 12–23 per cent of farm dogs were carriers. Although the number of associated human cases was small—only 214 in nine years—clearly more attention should be given to the routine treatment of dogs, and prevention of dog infection by the disposal of sheep carcasses, and a recent pilot scheme initiated by the British Veterinary Association in Wales has shown that this can be highly effective.

Liver fluke is a menace in many parts of Britain in wet years although infections are most serious in the higher rainfall areas. It has a two-host life cycle, the adult being found in the bile ducts of the liver in sheep, cattle and other animals, and the young stages in a snail. This snail only colonises areas where the soil is fully saturated with moisture for considerable periods, but is not found in permanent ponds and streams. In dry periods and during the winter, it burrows in the mud and can survive inactive for a considerable time.

Hatching and development of fluke eggs is restricted by their temperature and moisture requirements to the period

May to September, coinciding with the season of activity of the snail. The hatched larva must find and penetrate a snail within about twenty-four hours. Within the snail a number of development stages occur, the final stage leaving the snail and attaching to vegetation, forming a protective capsule. This stage is infective to the sheep and, if eaten, it is liberated in the intestine, burrows into the abdomen and attacks the liver, where it wanders in the tissue for eight to ten weeks, then settles in the bile ducts.

The snail population builds up steadily through the late spring and summer, and the number of infected snails increases proportionally. The development of the parasite within the snail takes about three months so that pasture infection from the snail begins in midsummer and reaches a peak in September–October, declining again as snails die off in the winter. Acute fluke disease due to liver breakdown caused by large numbers of young, migrating flukes is therefore common in early autumn, and the chronic disease occurs rather later as the adults cause anaemia and interfere with liver function.

The economic importance of fluke infection varies from year to year, depending largely on rainfall, but some degree of continuing control is essential in many areas, and this may be approached in several different ways. The most effective approach is to get rid of the snails by removing the wet habitats which they require through drainage schemes, or where the suitable areas are small they can be fenced off and grazing prevented or limited to safe periods. Snail destruction by chemicals appeared more promising in recent years with the development of safe and more effective preparations, but their use has been limited by cost and the problem of how to apply them over extensive boggy areas.

The traditional method has been to attack the flukes in the sheep with drugs, and in the past this was only moderately successful since the common ones like carbon tetrachloride were only effective against adult flukes and not against the migrating liver stages. This meant they did not prevent the acute disease, and only removed a proportion of the infection at any one time. However, the modern drugs are much more effective, and remove most of the migrating stages as well as virtually all the adults. They are therefore effective in controlling acute outbreaks as well as the chronic condition, but in addition they offer the possibility of achieving a long-term reduction of farm infection to a low and much less dangerous level. The usual practice has been to dose sheep several times from autumn onwards, when adult fluke numbers are highest, to prevent deaths and severe loss of condition, but this does not prevent a significant degree of infection becoming established and therefore a considerable output of fluke eggs on the pasture. In the preventive approach (pioneered by the Hill Farming Research Organisation in Scotland) treatment is started in June and five doses are given at about six-week intervals between June and the end of January—in hill flocks treatment can be timed to coincide with routine gatherings. This rate of treatment, although more expensive, prevents any significant numbers of adult flukes developing and greatly reduces pasture contamination with fluke eggs, thereby both improving productivity and minimising the risk of severe disease in bad fluke years, and it may well be possible to follow a three-year preventative programme by a similar period with no treatment at all. Alternatively, the frequency and timing of treatment can be adjusted according to the Ministry of Agriculture's 'fluke forecast', which predicts the severity and time of onset of infection, based on a knowledge of the rainfall levels required to maintain suitable conditions for the snails. Cattle are equally suitable hosts, though usually much less heavily infected than sheep, and fluke infection can have a marked effect on milk yield. It is therefore essential to treat cattle as well as sheep in a comprehensive fluke control programme.

Roundworms cause parasitic gastro-enteritis, thought to be economically the most serious disease of sheep, but equally important is the annual wastage of production caused by the moderate worm burden of healthy sheep, for even a light infestation can cause a seasonal liveweight gain 25 per cent lower than that of worm-free lambs. In worm control attention is largely directed to animals in the first year of life, as this is the age group most affected by roundworms, older animals developing a degree of resistance to infection.

Fortunately in recent years we have made a lot of progress in producing highly effective worm remedies, and in understanding the pattern of infection with these parasites. Treatment with modern broad-spectrum drugs produces virtually worm-free sheep and correct pasture management can produce almost worm-free grazing, so that by a combination of dosing and moving sheep to clean pasture at strategic times of the year we can achieve a high degree of parasite control.

The common worm life cycle (see Figure 6) starts with adults in the gut, and egg laying by the females. The eggs are passed out in the dung of the sheep, and require warmth and moisture to develop, so that eggs deposited in winter are killed off. Under the right conditions, generally March to early October, a small larva hatches from the egg, grows and feeds on bacteria in the dung. After casting its skin twice the larva becomes the non-feeding but resistant infective stage, which migrates from the dung into the grass, where it is picked up by the grazing animal. Only this twice-moulted infective stage is capable of further development in the sheep, the earlier stages being destroyed if eaten.

There are two key features of the pasture stages of the life cycle. The first is that development from egg to infective stage

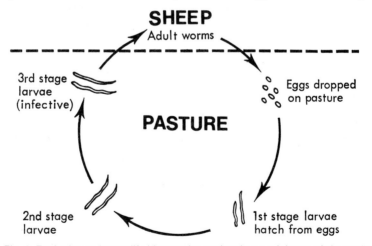

Fig. 6. *Typical roundworm life history; the cycle takes a minimum of six weeks and allows several generations in each grazing season.*

is slow in spring, taking up to ten to twelve weeks, but gets faster as the ground warms up and only takes two to three weeks in midsummer. This means that eggs dropped in April and eggs passed in June are translated into infective larvae at much the same time, and the main wave of infection on the grass occurs in July–August. The second is that once infective larvae are present on pasture they can survive a long time, particularly over winter. In this way pasture grazed by infected sheep one summer remains infective until the following June. The major exception to this sequence is seen in small intestine parasites of the Nematodirus group (Figure 7). In these the egg develops extremely slowly, and the pre-infective stages occur within the egg shell, taking at least six months to reach the infective stage, so that this generally takes until the autumn. Due to the low autumn temperatures, most of the infective stages usually remain dormant in the egg shell and only emerge the following spring when the soil warms up. This is virtually a twelve-month life cycle and infection is generally passed directly from one year's lambs to the next.

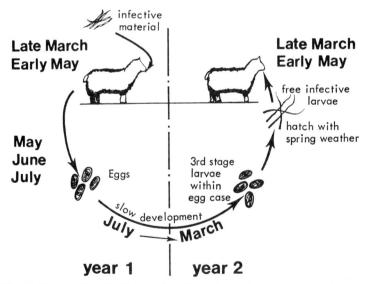

Fig. 7. Nematodirus life history; the life cycle requires approximately twelve months to give one generation per year.

Taken in by the host in spring or summer grazing, the infective larva passes to the stomach or intestine, and burrows into the gut wall. Two weeks later it returns to the surface, becomes adult and the female begins egg laying a week later. However, in late autumn and winter the infective stage of the commonest and most important stomach worm invades the gut wall and stays there in a hibernating condition. In spring, particularly in the pregnant ewe, it becomes active again, emerges and matures to lay eggs, and this is now thought to be largely responsible for the great increase in worm egg output of ewes in the two to three months after lambing. We may conclude that infection from midsummer to early autumn is most important in lambs, developing rapidly and causing anything from a severe check in weight gain to obvious disease and death. Late autumn–winter infection is most important to ewes, since it will be the source of next year's eggs to start the pasture contamination cycle again, whereas most of the lambs will be slaughtered while the larvae are still quiescent.

The seasonal pattern of infection can be divided into two phases, pre-weaning and post-weaning. Pre-weaning infection occurs when ewes and lambs are turned in spring on to pasture which carried sheep the previous year. Overwintered larvae are picked up by the lambs as soon as they begin to graze, and a quite substantial worm burden can develop in May–June which can cause a serious check. The most important parasite at this time is Nematodirus, large numbers of dormant larvae suddenly hatching and causing massive infections, being most dangerous in a late spring when hatching is delayed until the lambs are eating substantial amounts of grass. In addition, the ewes will also pick up overwintered larvae, increasing their worm burden and leading to greater spring egg output (see Figure 8).

Post-weaning infection occurs from about mid-July onwards when the lambs are exposed to the main wave of pasture larval activity resulting from eggs deposited since spring, particularly from the increased egg output of lactating ewes. Large numbers of larvae are usually taken in, since the lambs are now weaned and entirely dependent on grass, and a severe check or clinical disease is common from August

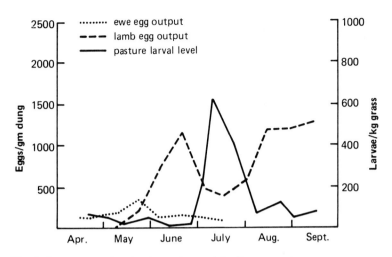

Fig. 8. Sources and levels of roundworm infection on pasture.

onwards. Eggs passed at this time develop into infective larvae which are either eaten later in the autumn or overwinter to present a threat for the next year.

Having clarified the main sources of infection and risk periods, it becomes possible to organise control systems. Obviously, in view of the long development period and long survival of pasture infection, short-term rotational grazing has no value in worm control, and equally, routine monthly dosing is quite uneconomic, since worm challenge occurs at specific times and treatment at other times is wasted. The sensible approach is to apply strategic control measures at the critical stages of the season, and the most important factor is the provision of clean grazing for dosed animals.

The first risk period coincides with lambing and the early suckling period when ewes and lambs are turned on to fresh spring pasture. If this grazing is free of infective larvae, then no pre-weaning infection of lambs or ewes will occur. Since overwintered infection dies out by June, this clean pasture need only suffice for the first couple of months, and in June any field may be considered clean if it has been rested from sheep since February. What is absolutely essential is to avoid land which carried lambs the previous May–June in order to prevent Nematodirus infection.

The major risk, however, is the post-weaning infection from the ewes and this can be avoided in two ways. It can be tackled at source by dosing the ewes at or just after lambing—a convenient time is when batches of ewes and lambs go out from the lambing shed or paddock. With modern drugs the spring flush of ewe egg output is almost entirely suppressed, and pasture contamination is so reduced that minimum lamb infection results even if they graze the same pasture throughout the summer. However, it is essential that dosed ewes should go on to clean grazing, otherwise they will immediately become reinfected and continue to pass eggs. Alternatively, since we know that spring-contaminated pasture does not become dangerous until July, infection can be avoided by moving the lambs by mid-July, which conveniently coincides with weaning in many lowland flocks. The

PLATE 9
Lambs, drenched with thiabendazole at weaning, grazing clean pasture.

Photo University of Newcastle

safest system combines both practices, i.e. the ewes are dosed on to clean pasture at lambing, and lambs are dosed and moved to clean pasture again at weaning (see Plate 9). These two strategic actions give a very high level of worm control, and by preventing contamination they increase the availability of clean pasture, so that the clean grazing system becomes progressively easier to operate.

Clean grazing is really essential to efficient worm control, so it is necessary to know how it may be provided, since this is the limiting factor in many farms, and the major obstacle to increased productivity on hill farms. It is available in three main ways. First and most useful, the dominant sheep parasites are not transferable to cattle and vice versa, so that pasture grazed by cattle one year is clean for sheep the next. Secondly, pasture conserved as hay or silage is generally rested from grazing in spring and the post-conservation growth is then clean. Thirdly, on arable farms clean pasture is available from grass grown as a break crop after cereals. Following these principles, a suitable system to provide clean grazing at critical times can be designed for almost any farm, and agricultural advisers are usually willing to assist with this. However, even where clean grazing is extremely limited or unavailable, a knowledge of these principles allows dosing to be concentrated at times of maximum risk for maximum effect, namely in spring for the ewes and midsummer for lambs.

Lungworms in sheep appear to be fairly widely distributed, but there is not much information on their economic importance. They are again mainly a problem of the lamb crop, but infection develops rather later than the gut roundworm burden, producing coughing and occasional clinical disease in late autumn. Since lungworm infection tends to be sporadic, and particularly associated with wet autumn conditions, specific preventive recommendations cannot be given, but modern worm drugs are effective against lungworms as well as gut roundworms and therefore have dual activity when used for strategic medication.

The search continues for drugs with a wider and wider range of activity to give more comprehensive parasite control by strategic medication. Combined activity against lung and gut roundworms and tapeworms is a common property of the benzimidazole group of drugs, and one of these, albendazole, also has a useful degree of activity against liver-fluke. The new avermectins, already mentioned, show that combined activity against external and internal parasites is possible. At the same time more information is becoming available on the seasonal patterns of parasite infection and the factors influencing the size of the infection. As a result, control programmes are becoming increasingly effective and we can continue to look forward to further developments in this important field of preventive medicine.

APPENDICES

Appendix I

SHEEP-MEAT IN SOME EEC COUNTRIES

	1970–4	1978	1979
Italy			
Home production (000 tonne)	33	37	37
Net imports (000 tonne)	27	30	31
Consumption (000 tonne)	60	67	68
Consumption/head (kg)	1·1	1·2	1·2
Self-sufficiency (%)	55·0	55·2	54·4*
Germany			
Home production (000 tonne)	14	18	21*
Net imports (000 tonne)	5	28	31
Consumption (000 tonne)	19	46	52
Consumption/head (kg)	0·3	0·8	0·9
Self-sufficiency	73·7	39·1	35·8*
Belgium–Luxembourg			
Home production (000 tonne)	3	4	4
Net imports (000 tonne)	8	16	16
Consumption (000 tonne)	11	20	20
Consumption/head (kg)	1·1	2·0	2·0
Self-sufficiency (%)	27·3	20·0	20·0
France			
Home production (000 tonne)	128	147	149*
Net imports (000 tonne)	42	55	60*
Consumption (000 tonne)	170	202	209
Consumption/head (kg)	3·3	3·8	3·9
Self-sufficiency (%)	75·3	72·8	71·3*

* Estimates

Appendix II

BODY CONDITION SCORING*

The success of the system depends primarily on the clear and precise description of certain physical characteristics identifiable in sheep of different degrees of fatness. These characteristics are assessed in the lumbar region (on and around the backbone in the loin area immediately behind the last rib) and above the kidneys.

The first stage is the assessment of the prominence (the degree of sharpness or roundness) of the spinous processes of the lumbar vertebrae—these are the bony points or spikes rising upwards from the backbone. The second step is to assess the prominence of, and the degree of fat cover over, the transverse processes of the vertebrae—these are the horizontal bones coming out from either side of the backbone. Thirdly, the extent of the muscular and fatty tissues below the transverse processes is judged by the ease with which the fingers pass under the ends of these bones. Finally, the fullness of the eye muscle area, and its degree of fat cover, in the angle between the spinous and transverse processes, is estimated. Animals are then awarded a score according to the following scale:

SCORE 0. *Extremely emaciated and on the point of death. It is not possible to detect any muscular or fatty tissue between the skin and the bone.*

SCORE 1. *The spinous processes are felt to be prominent and sharp. The transverse processes are also sharp and the fingers pass easily under the ends, and it is possible to feel between each process. The eye muscle areas are shallow with no fat cover.*

SCORE 2. *The spinous processes still feel prominent, but smooth, and individual processes can be felt only as fine corrugations. The transverse processes are smooth and rounded and it is possible to pass the fingers under the ends with little pressure. The eye muscle areas are of moderate depth, but have little fat cover.*

SCORE 3. *The spinous processes are detected only as small elevations; they are smooth and rounded, and individual bones can be felt only with pressure. The transverse processes are smooth and well covered, and firm pressure is required to feel over the ends. The eye muscle areas are full, and have a moderate degree of fat cover.*

SCORE 4. *The spinous processes can just be detected, with pressure, as a hard line between the fat-covered eye muscle areas. The ends of the transverse processes cannot be felt. The eye muscle areas are full, and have a thick covering of fat.*

SCORE 5. *The spinous processes cannot be detected even with firm pressure, and there is a depression between the layers of fat in the position where the spinous processes would normally be felt. The transverse processes cannot be detected. The eye muscle areas are very full with very thick fat cover. There may be large deposits of fat over the rump and tail.*

In any flock it is unlikely that condition will range over more than two scores at any one time, or more than three scores over the whole year. In practice, therefore, the range of fatness is narrower than that encompassed by the scoring system, and the scores 0 to 5, as set out above, become inadequate to describe variations within a flock. In this situation it becomes almost automatic to assess sheep to the nearest half score.

---

* From MLC Publication, *Feeding the Ewe*

Appendix III

COMPOSITION OF MAIN FEEDS*

	Dry matter g/kg fresh matter	Metabol- isable energy MJ/kg DM	Crude protein g/kg DM	Degradability of protein (%)
Grains				
Oats	860	12·6	105	85–95
Barley	860	13·0	120	85–95
Maize	860	13·5	98	60–70
Wheat	860	13·9	125	70–80
High protein supplements				
Soya bean meal (extracted)	900	12·3	503	45–65
Groundnut meal (decorticated)	900	11·7	504	65–85
Fish meal	900	11·1	701	25–45
Roots				
Mangels	125	12·4	89 ⎱	85–95
Swedes	120	12·8	108 ⎰	
Hay				
Clover	850	9·6	180	35–65
Grass:				
Leafy (very good)	850	10·1	132	
Early flower (good)	850	9·0	101	
Seed set (moderate)	850	8·4	85	70–80
Stemmy (poor)	850	7·5	92	
Barley straw	860	7·3	38	80
Grass silage				
Leafy (very good)	250	11·8	159	untreated 80–90
Early flower (good)	250	10·5	145	formic acid added 30–50
Full flower (moderate)	250	9·2	132	

* From MLC Publication, *Feeding the Ewe*

Appendix IV

SPECIMEN CONCENTRATE MIXTURES*

| | | | Ingredients (per cent) | | | | | |
Oats	Barley	Maize	Wheat	Dried sugar beet pulp	Beans and peas	Soya bean meal	Ground-nut meal	Fish meal
Lowland ewes: pregnancy and lactation								
40	30	20	—	—	—	—	—	10
25	30	20	—	—	—	15	10	—
—	72	13	—	—	—	—	7	8
15	40	—	15	10	—	15	—	5
20	25	8	—	10	25	—	7	5
30	30	—	20	—	—	20	—	—
Hill ewes								
—	25	50	10	—	—	—	7	8
23	—	40	20	—	—	12	—	5

Notes:
1. The required nutritive value of the concentrate ration will depend very largely on the quality of the roughage fed.
2. Include vitamin/mineral supplement at the rate of 2·5 per cent in the concentrate ration, i.e. 25 kg per tonne of mix.
3. The levels of groundnut meal in the example rations must not be exceeded.

* From MLC Publication, *Feeding the Ewe*

INDEX

Abortion, causes of, 142
Adaptation of hill breeds, 29
Age of ewe replacements, 73
Alternate husbandry, 35
Animal Breeding and Research
   Organisation, 44, 125
Antiserum, 102, 155
Arable sheep farming, 28
Australian sheep farming, 16

Bakewell, Robert, 26
Benzimidazoles, 178
Black disease, 153
Bladderworms, 175
Blowfly, 109
Bluefaced Leicester, 42
Border disease, 144
Border Leicester, 27, 42
Braxy, 152
Breeding behaviour, 78
Breeding plans, 117
Breeds and their functions, 39

Carpet wools, 22
Castration, 81, 147
Cattle and sheep, 29
Charollais, 40, 126
Cheviots, 46
Choice of ewes, 71
Clean grazing, 36, 83, 183
Clostridial diseases, 151
Clover, for fattening, 102
Clun Forest, 39, 116
Clun Forest breeding, Cockle Park, 119
Coccidiosis, 150
Coke of Norfolk, 25
Colbred, 35, 42, 124
Colibacillosis (scours), 147
Concentrate feeding, 63, 80, 105
Condition scoring, 57, 186 (Appendix
   II)

Copper deficiency (swayback), 148
Copper toxicity, 55, 149
Corriedale, 14, 123
Creep grazing, 85
Cross Suffolk, 41, 51

Demand for sheep meat, 18
Dipping, 169
Disease classification, 136
Disease diagnosis, 137
Docking, 81, 153
Dorset Down, 40
Dorset Horn, 42, 47
Dosing for worms, 182
Down breeds, origins, 27
Drysdale, 123

Economic performance factors, 130
Enclosures, 25
Energy and protein, 59
Enterotoxaemia, 152
Enzootic abortion (kebbing), 143
Ewe breeds, 49
Ewe condition scoring, 58, 186
   (Appendix II)
Ewe depreciation, 73
Ewe replacements, 70
External parasites, 167

Feed blocks, 67
Feeding in lactation, 64
Feeding in pregnancy, 62
Finishing store lambs, 101, 104
Finnish Landrace, 48
Fixed costs, 127
Flock maintenance, 73
Fluke forecast, 177
Footbath, 159
Footrot, 157
Forage crops, 101

Gas gangrene, 152
Gid, 175
Grass sheep, 28
Grass tetany (hypomagnesaemia) 141
Grazing, creep, 86
Grazing, quality of, 83
Grazing, rotational, 84
Grazing, set, 84
Greyface, 43, 51, 71
Gross margins, 127
Growth of foetus, 62
Growth of lambs, 46

Hampshire, 40
Haybox, 96
Headfly, 170
Hill breeds, 44
Hill Farming Research Organisation,
    66, 177
Hoggets, 104
Housing, 90
Hydatid cyst, 175
Hypocalcaemia, 140
Hypomagnesaemia, 141

Ile de France, 40
Improvement of crossing breeds, 121
Improvement of ewe breeds, 120
Indoor finishing, lambs, 106
Insecticides, 169
Internal parasites, 174
Inwintering, flock management, 96
Inwintering, pros and cons, 91, 94

Jacob sheep, 47
Joint-ill, 147

Kebbing (enzootic abortion), 143
Keds, 168

Labour efficiency, 134
Lamb dysentery, 151
Lamb finishing, 101, 104
Lamb growth, 46
Lamb survival, 146
Lambing management, 80, 146
Lambing pens, 81
Lambing percentage and body
    condition, 58 (Table 8)
Lambing sickness (hypocalcaemia), 140
Lambing, timing of, 77
Lice, 168
Liver fluke, 175

Lleyn, 47, 72
Long keep stores, 104
Longwool crossing breeds, 41
Louping ill, 173
Lungworms, 183

Masham, 43
Mastitis, 144
Mating management, 78
Meat and Livestock Commission, 49,
    71, 128
Merino in Spain, 16
Merino in Australia, 17
Metritis, 144
Milkworm, 174
Milk Production, 17
Milk yield and lamb growth, 46
Mineral supplementation, 142, 148, 166
Mites (scab), 168
Mules, 43
Multiple births, 51 (Table 7), 82
Mutton and lamb, demand for, 19
Mutton and lamb production, 18

Navel-ill, 147
Nematodirus, 179
New Zealand sheep farming, 14
New Zealand 'Sheeplan', 114
Nitrogen usage, 131
Nutrition of ewes, 60, 62

Oldenburg, 44
Orf, 162
Oxford, 40

Paddock mating, 79
Pasture, clean, 83, 183
Perendale, 14, 123
Pining, 165
Pneumonia, 163
Pregnancy toxaemia (twin-lamb
    disease), 138
Prospects for expansion, 33
Protein, digestion of, 59

Raddling, 80
Ram-ewe ratios, 79
Rape, 102
Recording in Britain, 115
Refrigeration, 99
Romneys in England, 46, 53
Romneys in New zealand, 14, 46
Rotational grazing, 84

Roundworms, 177
Ruakura grazing trials, 84

Scab, 168
Scottish Blackface, 44
Scottish Halfbred, 43, 50
Scours, 147
Scrapie, 164
Selection objectives, 110
Selenium deficiency, 149
Sheep health programmes, 136
Sheep populations, 14
Short keep lambs, 101
Southdown, 27, 40
Spanish sheep farming, 15
Spraying, 169
Stocking intensity, 131
Stratification, 28
Stubble turnips, 103
Suffolk, 40, 122
Sugar beet tops, 104
Supplementary feeding, 63 (Table 9), 65, 67
Swaledale, 46
Swayback, 147
Swedes, 105
Synthetic breeds, 123

Tailing, 81

Tapeworms, 174
Teaser rams, 79
Teeswater, 44
Terminal sire breeds, 40
Tetanus, 153
Texel, 41, 126
Tick, 171
Tick pyaemia (cripples) 173
Tickborne diseases, 172
Tickborne fever, 172
Townshead, Lord, 25
Transhumance, 15
Twin-lamb disease, 138

Upland sheep breeds, 45 (Table 5)

Vaccination, 155, 162, 163, 173
Vasectomy, 79
Ventilation, importance of, 91, 163

Watery mouth, 147
Weaning, 87
Welsh Halfbred, 43, 71
Welsh Mountain, 44
Wensleydale, 43
White muscle disease, 149
Wool, characteristics of, 52
Wool, importance of, 20
Wool and synthetic fibres, 22
Wool, uses of, 53